This book should be returned
stamped below unless an extension
Application for renewal may be
Fines at the approved rate will
overdue by a week or part of a week

CW01090608

SCSU Library
Alnwick Castle
Alnwick England

CANCELLED

NORTHUMBERLAND COUNTY LIBRARY

MORPETH

S534567

J 595.796

The World of an Ant Hill

other titles in this series

THE WORLD OF AN ESTUARY
Heather Angel

THE WORLD OF A STREAM
Heather Angel

THE WORLD OF A TREE
Arnold Darlington

THE WORLD OF AN ISLAND
Philip Coxon

THE WORLD OF A MOUNTAIN
William Condry

THE WORLD OF A BEEHIVE
John Powell

THE WORLD OF A HEDGE
Terry Jennings

THE WORLD OF THE CHANGING COASTLINE
Jill Eddison

THE WORLD OF AN
Ant Hill

M. V. Brian

with line drawings by
Wilhelmina Mary Guymer

Faber & Faber
London · Boston

First published in 1979
by Faber and Faber Limited
3 Queen Square London WC1
Printed in Great Britain by
BAS Printers Limited, Over Wallop, Hampshire
All rights reserved

© *1979 M. V. Brian*
drawings © *1979 Faber & Faber Ltd.*

British Library Cataloguing in Publication Data

Brian, Michael Vaughan
   World of an ant hill.
    1. Ants—Behaviour—Juvenile literature
    2. Social behaviour in animals—Juvenile literature
   I. Title
595.7′96′045   QL568.F7

ISBN 0-571-11323-0

Contents

Illustrations	7
Acknowledgements	9
1. Different Kinds of Ant Hill	11
2. Ants as Social Insects	20
3. Hunting and Scavenging	29
4. The Collection of Plant Juices and Seeds	34
5. The Care of Young Ants	41
6. Nests with Ants under Observation	48
7. Ant Hills as Homes	55
8. Ant Hills through the Year	60
9. Winged Ants and Nuptial Flights	64
10. The Foundation of Colonies and their Growth	69
11. The Pattern of Ant Communities	74
12. Plants and Animals that use Ant Hills	80
Further Reading	90
Index	91

Illustrations

PLATES

1. Yellow ant mounds in grassland area — 12
2. Closer view of Yellow ant mounds — 13
3. Wood ant mound in open pine forest — 14
4. Wood ant mound: close-up view — 15
5. Worker ants carrying pine needles on Wood ant mound — 17
6. Colony of Driver ants resting for the night — 18
7. Aphids and a Wood ant worker — 22
8. Aphids tended by Black garden ant — 24
9. Wood ant combing its left antenna — 25
10. Wood ants taking prey to their nest — 30
11. Wood ant feeding another worker — 35
12. Black garden ant collecting honey-dew from an aphid — 36
13. Aphid cluster attended by two Wood ant workers — 36
14. Wood ants tending aphids — 38
15. Fungus ants taking a piece of leaf back to the nest — 39
16. A Tropical ant: brood in various stages — 42
17. Black garden ant with cocoons — 45
18. Black garden ant getting a grip on a cocoon — 46

ILLUSTRATIONS

19. Black garden ant throwing away a dead insect	47
20. Male Wood ant ready for nuptial flight	65
21. Queen ant of fungus-culturing species	70
22. Yellow ant hill	81
23. Yellow ant hill with soil scratched by rabbit	81
24. Material of Wood ant hill scattered during winter	82

FIGURES

1. Wood ant worker	21
2. Head of Wood ant worker	22
3. Red ant and Wood ant, vertical sections	23
4. Black garden ant and Turf ant	27
5. Brood stages of ants	43
6. Window on to an ants' nest	49
7. Two types of artificial nest	51
8. Sucking tube for collecting ants	53
9. Graph showing temperature in a Wood ants' nest	54
10. Nests and foraging tracks of Wood ant colonies	58
11. Larva of the beetle *Atemeles*	84
12. Adult of *Atemeles* and Red ant host	85
13. Caterpillars of the Large blue butterfly and ants	86

Acknowledgements

Plates 5, 7–14 and 17–20 were taken by S. L. Mason and Plates 6, 15, 16 and 21 by D. J. Stradling. The remaining photographs are by the author.
For the diagrams, the following sources have been used:

Figure 1a: W. S. Creighton. *Bulletin of the Museum of Comparative Zoology*, Harvard College, Volume 104, Cambridge, Mass. U.S.A. 1950.

Figure 1b: K. Gosswald. *Die Rote Waldameise im Dienste der Waldhygiene*. Metta-Kinau, Lüneburg 1951.

Figure 3a: C. Janet. *Anatomie du Gaster de la Myrmica rubra*. Carré et C. Naud, Paris, 1902.

Figure 3b: G. Dlussky. *Ants of the Order Hymenoptera*. Published Moscow 1967.

Figure 4: W. S. Creighton. As figure 1.

Figures 9 and 10: A. Raignier. *L'économie thermique d'une colonie polycalique de la Fourmi des bois*. *La Cellule* 51, 281–368 (1948).

Figures 11 and 12: B. Hölldobler. *Scientific American*, March 1971, pages 86–93. From drawings by Mrs. T. Hölldobler.

Figure 13: Drawn from photographs taken by Dr. J. Thomas.

1. Different Kinds of Ant Hill

Ant hills are the homes in which ants spend most of their lives. They only leave them rather late in life to collect food and nest material, and before they go far they make sure they know how to find their way back. In this country nearly all ants make their homes (nests) in soil and they do not always bother (or manage) to pile the soil up into a hill. They may even take over and enlarge small hills started in other ways; for example, many grasses and some mosses form tussocks that are often used as a foundation for ant hills. Moles as a result of their tunnelling underground throw up heaps of soil that are at first loose and crumbly but after a year or two flatten and get a growth of small grasses and herbs on top. At this stage, ants often move in and start cutting out a nest for themselves under the top crust of soil. Again, when trees are cut down a stump about 10 cm high is usually left. At first this is much too hard for ants to bite into, but in a few years the bark begins to separate away and later on the outer layers of true wood begin to soften and rot. Many ants find the south sunny sides of these stumps ideal for cutting out a nest. Rock outcrops and stones are in a sense also hillocks and against the sides of these ants often build nests. They also excavate underneath stone slabs where old walls have crumbled or under slates or fallen house tiles, or paving stones in gardens and cities. You have only to poke your finger into a hillock to see whether ants live there or not; if they do, then hundreds will soon come out to explore the damage and attack the cause if it is still there.

The simplest nests are just sets of rooms or chambers, connected by holes, by vertical shafts, or by horizontal galleries. The chambers vary in size but as a rule, the bigger the ant, the bigger the chamber. They are not regular like the cells of bee or wasp combs, but are big enough to hold many ants and their brood at one time: usually they are designed so that the height of the ceiling just allows an ant standing on the floor to reach up and walk on it. In fact, ants walk on the ceiling as much as on the floor, for this is usually cluttered with brood and food, and reaching down from the ceiling is a much easier way of dealing with young ants than creeping over them. Each chamber is mudded with soft moist soil so that there are no crevices.

There are two really skilled ant builders in this country. One is the Yellow ant (*Lasius flavus*), which lives in pastures, and the other is the Wood ant (*Formica rufa*), which lives in open forests. Yellow ants are responsible for the impressive scenery of hillocks which cover the whole of some hillsides in chalk and limestone downland. These hills

1. In undisturbed areas of grassland the Yellow ant can make very many mounds of soil which include grasses and herbs, both inside as roots and on top as shoots. This is a photograph of an old pasture of chalk in Dorset. Cattle, sheep and rabbits graze the herbage.

DIFFERENT KINDS OF ANT HILL

were often called Emmet Hill; emmet is an old word for ant. The hillocks are only a few metres apart and each is up to half a metre high and a metre in diameter at its base. The base is circular only in very new nests; as they grow, they become more and more oval in ground plan and more distinctly flat-topped. They are made of soil, but this is quickly covered by a growth of mosses, grasses and a variety of herbs.

2. Closer approaches to these mounds show how much more steep-sided and flat-topped they are than the smoothly curved dome-shaped Wood ant mounds. The flowers that live on them are often quite different from those in the surrounding vegetation.

Yellow ants have very strong jaws which unlike ours work sideways and can be used to cut soil particles from the deep layers (the sub-soil). They scrape the soil particles together into piles with their legs and carry them up to the top of the new growing surface. The soil that is brought up, because it comes from a layer not normally exposed, often has quite a different colour from the rest. Once on the surface, provided there is moisture either from rain or dew, the particles are stuck on to the top of the existing soil and built into a set of chambers by the ants. This is done largely from the inside, so that they rarely appear in daylight. Each chamber connects with a neighbouring chamber by a very small hole or a longer passage, very little wider than the width of one ant. Yellow ants welcome plant growth because the roots strengthen their nest; sometimes they start nests against stones and rocks, for support, but later cover these over.

Each year for the first few years the workers raise the height of the mound by about 10 cm. Later less is built on to the nest, and allowing for loss due to winds and rain blowing and washing the loose soil away, it is not surprising that the formation of a good-sized hill may take twenty years or more. Though Yellow ants rarely come into the daylight and you rarely see them unless you break the surface, only a centimetre below the top crust there is a silent and energetic throng. A whole hillside covered with mounds of this kind represents an enormous activity on the part of these soil insects.

3. A mound of the Wood ant in open pine forest; the sunshine is warming the south side and you can see how, by piling dead vegetable matter up, the ants have managed to take their nest into the sunlight above quite tall grass and heather.

DIFFERENT KINDS OF ANT HILL

The other great mound builder in this country, the Wood ant, lives, as I have said, in open forest. The Wood ant prefers forest which has a few big mature trees and shrubs, with glades of grass and herbs in between. In the south of England Wood ants are usually found in sandy well-drained heath-carrying soils, but in Wales and the north and in Scotland they can be found in steep hillside forests, mainly of oaks; they also occur in open moorland. Wood ant nests are unmistakable, not only because the ants are conspicuous large brown insects about a centimetre long, far bigger than any other British ant, but because they can be seen running all over the surface. The nests are mounds made of vegetable debris such as pine needles, dead twigs, dead leaf stalks and all kinds of litter from trees and bushes.

4. Another Wood ant mound: notice the dome-like shape and the fact that many pine needles are used in its construction. Here most of them are lying up and down hill (like thatch) and this must enable rain to be shed effectively.

Though many are no bigger than the hills of the Yellow ant, some reach a metre in height. At all sizes and ages their base is much more nearly circular than that of the Yellow ant mound, and it may be more than a metre across. The general shape is between that of a dome and a cone. The top is slightly flattened, though this, as you will see later, varies with the climate and with the type of Wood ant. Most ant hills are in sunny spots, but a few will be found deeper in the shade of the wood. Be careful how you approach these nests; the ants can see you from a little distance and will stand with their hind bodies brought forward between their legs in such a way that they can, with the slightest provocation, squirt a jet of acid (naturally enough called formic acid, since *formica* is the Latin for 'ant'). You will also find that as you approach the mound many streams of ant traffic connect with it, and without realising it you may tread on one. Worker ants in these streams run vigorously both ways; some are coming back to the nest laden with booty, others are going out to nearby or even distant trees, perhaps over a hundred metres away. This will be discussed in Chapter 3. If you see an ant track-way and investigate, you are bound to find a nest at one end or the other.

The Wood ant often starts on a tree stump and eventually buries it. Workers collect pieces of dead twig and leaf stalks from the surrounding area and are quite particular about the size; many pieces less than 10 cm long are rejected and so are many over 15 cm long. Each piece is taken as far up the growing hill as the ant can drag it, and then it is not placed in any particular direction (like the reeds on a thatched cottage, for example) but is left without any apparent relation to other nearby bits. The only condition is that it must be laid flat on the surface and not be sticking up at all. There is no doubt too that the ants try to fill grooves and hollows of any kind in the surface of the nest. They are more likely to drop their pieces of twig into a depression than anywhere else, with the result that the nest grows up in a smooth coneshaped way. Naturally the growth upwards is opposed by the general tendency for bits to fall downhill away from the top, and many ants spend a lot of time carrying them back up again. In addition to the smooth cone shape the dome has a great many holes in it that are opened and shut as the weather changes;

DIFFERENT KINDS OF ANT HILL

5. This photograph was taken near the surface of a Wood ant mound. You can see pine needles, and worker ants grasping them near their ends. In this way they carry them from lower to higher levels. Notice how the antennae are stretched out in front of the head, the thorax with three pairs of legs and the large abdomen. The eyes of the worker on the left are particularly conspicuous.

these enable the inner atmosphere of the nest to be ventilated (see Chapter 6).

If Wood ants find a vertical object like a twig on the surface of their nest they will pull at it and flatten it out. If it is a living shoot they will gnaw the bark off it until it dies and then flatten it down. No plant shoots are allowed on these hills, as they spoil their shape. You can learn a lot about the building habits of Wood ants by dropping various things of different sizes and shapes on to their nest dome and seeing what they do with them. Do not handle the objects too much beforehand, however, as the ants do not like human smell.

The Yellow ants and the Wood ants are unusual in making huge, unmistakable and conspicuous hills. Most ants simply dig nests in soil, or cut them out of the dead wood of fallen branches and tree stumps. A common ant, like the Yellow in shape, but black, hairy and bigger, is often found in parks, gardens and scrub and is called the Black garden ant (*Lasius niger*). It excavates a lot of soil, taking it from under slabs and stones and throwing it out on top and around. Much of the soil is just washed or blown away, but if it stays, ants will make a nest in this too. So they form a hill nest without having special construction skill.

6. Though most ants build hills or excavate galleries in soft wood and soil, some, in the wet tropics, do not need to do so. Here you see a colony of Driver ants resting for the night under an old tree stump. In the morning they move off in columns and search for prey. There will be several hundred thousand workers in this cluster but only one queen.

There is another common group of ants that are all reddish brown and slender, and about 1 cm long. They may be found in parks, grassland and gardens, and are called Red ants (*Myrmica* species). They build by putting pieces of moist soil or litter against the stems and leaves of small herbs until a mound no bigger than an inverted cup is made. They also cut nests out of old tree stumps, rotting wood or soft soil. If you find a nest in summer, do not attempt to excavate it, as these ants sting quite fiercely, but mark its position and then go back in winter when the workers are torpid. Dig the soil out with a trowel and use a strong knife to cut any roots. Chambers and ducts can be more easily found and followed if you blow some French chalk in with a rubber puffer.

Only in tropical rain forests can some ants do without nests.

2. Ants as Social Insects

If you look at an ant closely you will probably be surprised by the delicacy of its structure: a great deal of equipment for life is packed into a very small space. You could really do with a hand lens for this or, ideally, a stereoscopic microscope magnifying ten times, but it is possible to see quite a lot without either. Perhaps the first time you examine a worker ant, it would be easier to kill it; simply drop it into warm water containing a little detergent and it will die immediately. Later, you will find no difficulty in holding an ant by the legs between your finger and thumb; you will not damage it as long as you are careful.

There are three main sections to its body: the head, the thorax and the abdomen. All insects are like this. The head carries a pair of antennae that are in almost constant use for groping and feeling in the darkness of the nest; surprisingly, they are used for smelling as well. Compared with most insects ant workers have only very small rigid, cellular eyes which can do little more than tell light from darkness, and distinguish vague shapes and movements. There are two jaws, each with a saw-like cutting edge; they are used for grasping, biting and chewing. The mouth is behind these and it has a tongue that can be pushed right out only when the jaws are apart. Various palps for tasting are fixed on either side of the mouth, which is of course the beginning of the digestive tube or gut. The antennae, palps and eyes are all connected by nerves to the brain, which is invisible inside the back of the head. It is a centre, like a computer in many ways, that

ANTS AS SOCIAL INSECTS

sorts, classifies and stores the coded messages that arrive down these nerves. After co-ordination the messages are referred to a built-in programme for action that has been tested and proved effective over millions of years. Any necessary adjustment, movement, action or other change is then arranged and controlled by the brain with its subordinate centres (called ganglia) which occur all the way down the body.

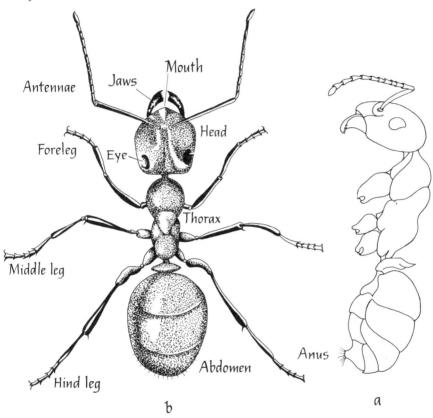

Fig. 1 Wood ant worker. a. side view b. top view.
Notice: head, thorax and abdomen; jointed antennae on head and three pairs of jointed legs attached to the thorax; toothed jaws and eyes on head; mouth in front between jaws and anus behind.

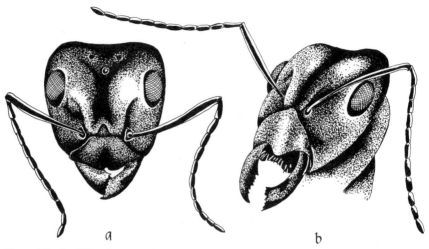

Fig. 2 Head of Wood ant worker. a. front or top view b. side view slightly oblique. Notice: eyes, antennae, jaws, upper lip.

7. This is a cluster of aphids feeding on the young shoots of an oak tree. They are of various sizes and ages and a Wood ant worker can be seen wandering over them and touching them with her antennae. Notice how her 'face' shows, below, the jaws held together, above, the roots of the antennae and higher up still, the eyes one on each side. The leg joints show well too.

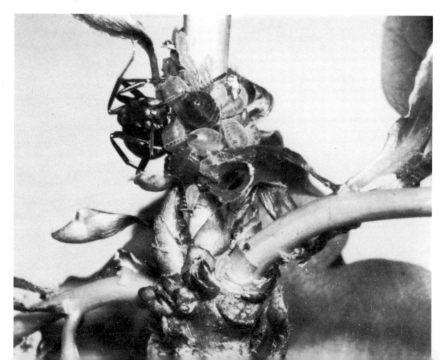

ANTS AS SOCIAL INSECTS

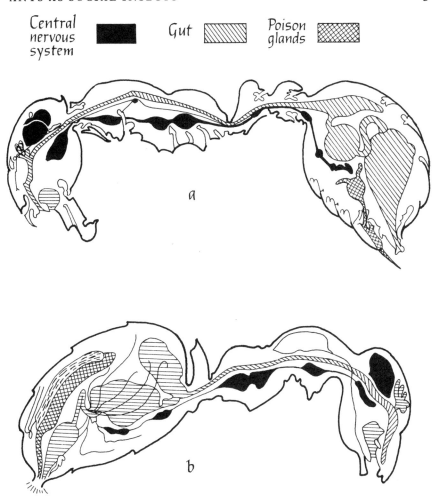

Fig. 3 a. Red ant, vertical section in the mid-line b. Wood ant, vertical section in the mid-line

These show only the gut, the poison glands and the central nervous system.

Notice: the crop for carrying liquids in the front of the abdomen followed by the digestive tubes ending in the anus. The head contains a pharynx that separates oil from water. The nervous system has a brain in the head that surrounds the gut and it has several ganglia down the ventral side of the thorax and abdomen. Poison is produced in glands in the hind abdomen, stored in a reservoir and then either squirted out (Wood ant) or forced down a sting (Red ant).

8. A huge cluster of aphids tended by a Black garden ant which is poising to clean its right antenna. It has its front right leg over the antenna and is running the comb down from base to tip. Particles of dust, plant spores and suchlike objects are pushed off or collected on the comb. Later the comb is washed by drawing it through the mouth.

The thorax is totally different. It is composed largely of muscles for moving the six jointed legs which end in claws; each first leg also carries a comb for cleaning dust off the antennae (plates 8 and 9). Most insects, in addition to three pairs of legs like this, have four wings coming out of the sides of the thorax, but worker ants are remarkable and exceptional in that they have no wings at all. With all these organs for movement attached, the thorax becomes a sort of engine for driving the insect body about.

The abdomen contains most of the digestive tubes which open by the anus near the tip to let out indigestible materials and waste products. It has storage depots to which the new food substances are carried and deposited until needed. The abdomen also contains the egg-tubes or ovaries; these are simple tubes down which young eggs move and grow until they are ripe. The ovaries join to form an egg-duct which opens just below the anus but is difficult to see, as it is covered by flaps of tough horny skin like that covering most of the body. The poison glands open here too.

ANTS AS SOCIAL INSECTS

9. Here a Wood ant has nearly finished combing its left antenna (look at the lower ant on the right). The comb of the other worker (on its left foreleg) shows well but the ant is apparently resting.

We have been looking at a worker ant. All workers are females, but there are also other females in the nest, called queens. Workers and queens are completely dependent on each other, since the work of the society is divided up so that each plays a different part. Generally speaking, the workers collect the food, defend the nest and tend the young, whereas the queens found new colonies, lay eggs and control worker behaviour. These names for the two types of female were given to ants a long time before their function in the society was understood, and it is very misleading to suppose that there is any likeness to our society. These mixed communities of females, composed of workers and queens, go on year after year all the year round. It is only by co-operating all the time in this way that ants manage to overcome the destructive and disturbing influences of other animals and plants, not to mention the rigours of seasonal change and the irregularity of weather.

The queens are bigger and more complicated than the workers. Young ones appear in summer and are noticeable because they have wings, two on each side of the thorax; but these are so hooked together that they beat as one. The thorax is nearly twice as big as that of the worker and able to hold the powerful extra muscles needed to drive the wings. The queen's abdomen is also bigger than that of the worker, as it holds an ovary containing many more egg-tubes. The head is much the same in both these types of female; if you have a lens you may be able to see a group of three very small eyes on the top of the queen's head, but there are few other differences. Males are present in nests at the same time as these young winged queens; they too have wings and are either much the same size or rather smaller and darker. As we shall see again in Chapter 9, both sexes fly out of the nest to a meeting place where they pair briefly. During mating, their abdomens interlock at the tip and the male transfers a package of sperm to the female. This is stored in a special pouch on the side of the egg-duct near where the eggs pass out. The sperm is held dormant in this pouch for the whole of the queen's life: perhaps for ten years. It is used during this time to fertilise eggs so that they may grow into young females, for ants (and bees and wasps) are unusual amongst animals in that unfertilised eggs, if they develop at all, grow into males. Once the males have paired they die and disintegrate or quickly become a prey to birds or other insects; their existence in the ant world never lasts long and they are never social.

The young queens, on the other hand, after pairing go and find new places in which to live and start new colonies. Once they have reached the right sort of place they tear off their wings along a line of weakness near their base. Queens with a life-time's store of sperm have no need to fly again, and, underground, wings may be a nuisance and get quite messy. This means that if you see big wingless females either alone or in small groups or in a colony, in spring or autumn, you will be looking at queens that have once had wings but broken them off; in fact, with a lens you can still see the old wing stubs. After ripping their wings off, the young queens start an intensive search for holes and crevices in the soil.

The narrow link between the abdomen and the thorax in the

ANTS AS SOCIAL INSECTS

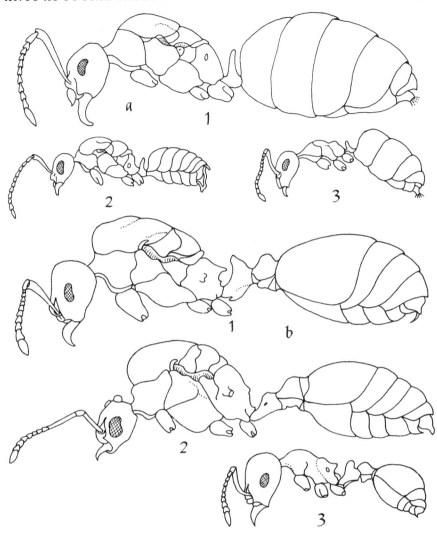

Fig. 4 Two castes of female and the males of the two main sub-families of ants. a. The Black garden ant representing Formicinae b. The Turf ant representing Myrmicinae.
The queens and males have wings that are not shown here, the workers do not. Only the bases of the legs are indicated.
Notice: two petiole segments for Myrmicinae and only one for Formicinae. 1, queen; 2, male; 3, worker.

female, whether worker or queen, is a feature of ants and also of bees and wasps. It gives great flexibility to the abdomen, so that the tip can be held in a variety of places and pointed in many different directions. This is a big advantage when laying an egg: a female can take the egg into her jaws as it comes out of the egg-tube opening. Since the poison apparatus also opens at the tip of the abdomen, it too can be brought round and a jet squirted forwards whilst the worker uses its eyes and antennae to control the aim. Ants which have stings instead of jets can manoeuvre their sting into a weak place in an adversary. These stings are no more than fine hollow needles connected to a reservoir of poison and they function like a hypodermic needle. They are used in exactly the same way: once the sting has been inserted, pressure is applied and a liquid poison forced into the intruder.

The link between thorax and abdomen is called the pedicel. Naturalists use it and the poison apparatus to distinguish between two big groups of ant species: one group has two joints in the pedicel and a sting, whilst the other group has only one joint in the pedicel and a poison jet. The former are said to belong to the sub-family Myrmicinae and the latter to the sub-family Formicinae. 'Sub-family' is used rather than 'family' because both these groups are classed together in the large family called the Formicidae (notice the 'd') to which all ants belong. The term 'family' as it is used here implies a general similarity, not an exact relationship, as it does when referring to human families. You will already have guessed from these names that the first group include *Myrmica* and the second *Formica*. *Lasius* is a formicine ant, not a myrmicine ant. There are many differences between these two groups besides the pedicel and the poison equipment.

3. Hunting and Scavenging

All ants have a very varied diet. This is a strong point in the struggle for a livelihood, as it enables them to use whatever food is, for the time, abundant and easy to find and collect. In this way a varied diet cushions them against unpredictable changes in food supply. All our species eat some sort of prey, but they will also pick up crumbs of cake, bread, cheese or fruit that they find lying about near picnic sites, or in parks or gardens. You may have noticed this.

Ants eat worms, centipedes, and spiders; they also eat insect larvae such as wireworms, caterpillars and grubs, and insect adults such as flies, aphids, moths and springtails. Such small animals do not all make easy game for ants; they are usually much quicker than the ant itself and they have an extraordinary number of ways of escaping. Some jump: these include springtails that live in and on the soil, froghoppers that suck plants and grasshoppers that eat many herbs and grasses; all, as their names imply, jump quickly out of reach if they detect an ant approaching. Many caterpillars, plant-bugs and beetles just drop off the plant they are feeding on, or dodge quickly round the side of the stem. Others, especially caterpillars and aphids, flick their bodies violently and make it very difficult for even persistent ants to grip them; beetles with their hard shiny surfaces are also very difficult to grip tightly. Beetles give out a pungent repellent substance as well; so do many other creatures: aphids exude a sticky wax from glands on their abdomen (called cornicles); cuckoo-spit insects make a disguising froth; and slugs and worms give out mucus.

10. A group of Wood ants round a small worm that they have chewed and are preparing to drag towards their nest, where it will serve as food for larvae. The worker at the bottom left has its jaws open preparatory to taking a pinch at the worm. Others are already biting and tearing and sucking up juices that they can squeeze out. The whole worm is then slowly dragged towards the nest-mound.

This is only a short list of the many ways and means that small animals use to avoid capture by ants.

Most species of ant hunt singly. Workers learn to recognise good places for hunting and they often tend to collect together simply as a result of this. They detect prey by its smell and its movement, and creep as close to it as they can, with their antennae stretched out in front, feeling and smelling. As they get near their intended quarry they fold their antennae back for safety, and with open jaws lunge suddenly forwards and snap their jaws together tightly. If they are skilful or lucky enough to get a grip on a leg, or something that sticks out, they hang on tight and try to tie their prey down and wear it out with struggling. Nearby ants are often attracted by the disturbance and come and take a grip too. Hunters may try to sting and squirt poison over their prey. Most of these encounters take place on the soil surface, on the leaves and twigs of trees, or even in the air spaces in soil.

If ants, or indeed any hunting animal were too clever at catching their prey they would eventually destroy the species completely and suffer themselves in turn. That is one reason why, in nature, it is usually only the weak and disabled that are caught and eaten. The well-formed, well-endowed, alert, vigorous and intelligent ones can escape and survive to keep the species going. The prey avoids destruction in this way, and maintains a good breeding stock which provides a regular food supply for its predators. The predator can even be thought of as a beneficial agent that prevents overcrowding of the prey by clearing up, neatly and effectively, all the superfluous individuals. It sounds strange, perhaps, but prey and predator are mutually useful.

Ants have no difficulty at all in catching insects that have just hatched from an egg, or emerged from a pupa (chrysalis) or have just changed their skin. At these times insects have soft bodies and can move only slowly and clumsily; certainly they cannot fly, for it takes quite a while for their wings to harden. Adult animals are often especially easy to catch whilst pairing, since the fact that they are closely interlocked nearly always interferes with their movements; one may even try to move in the opposite direction from the other. Damaged, lame or wounded animals are of course much easier to catch than normal ones; they may have wet wings that stick together, they may even be trapped in dew or raindrops on the leaves of plants, or more seriously they may have just been trampled and crushed by heavy beasts like cattle or horses or men. Many ants undoubtedly scavenge, that is, eat dead animals both small and large; people who prepare skeletons of birds and other animals often deliberately expose the corpses in places where it is expected that ants will come and clear the flesh away.

An ant that finds a manageable piece of food sets off home, or returns to the nearest permanent track-way first and then goes home. The food is either carried in the jaws, with the ant walking forwards, or dragged, with the ant walking backwards. This return journey is always remarkably straight, no matter how the ant takes its prey. Obstacles are avoided and the course is reset afterwards; it is a great navigational achievement. The work of many naturalists has shown that ants learn the area around their nest gradually; they take small trips at first and then longer and longer ones. This exploration teaches them the shape of their horizon; looked at from the ground even small bushes as well as trees and buildings make lumps of darkness against the sky and the foraging ants learn the pattern of these silhouettes. If you were to explore unknown territory without a map you would do the same, particularly at dusk or by moonlight when there is no colour.

In addition to this ability to learn shapes, foragers also have a more automatic sense of direction. This is less well understood, but it seems that ants have an internal record of their position in relation to their nest and the sun. There are two reasons why the sun is rather unreliable as a guide. One is obvious—it does not always shine. The other, perhaps less obvious, is that it moves in the sky 15° from east to west every hour. Both of these things the ants seem to be able to assess and compensate for automatically: the first, by picking up a pattern, invisible to us, from small patches of blue sky, the second by having an internal clock that measures the time from the start of a journey, that is, the time that they have been away from the nest. This is used to calculate how much the sun has moved during their trip. If they go out, say, in a line towards the sun and stay out three hours, they have to go back along a line with the sun three times 15°, that is 45°, to the west of them. On the return journey the sun would be shining from the side rather than from directly behind.

One more thing about ant wayfinding can be mentioned: foragers can tell whether they are going up or down hill. This applies especially whilst tree-walking; they can walk at a fixed angle to the vertical because their bodies have a way of sensing the pull of the earth (gravity) and the direction of this pull. Probably when they run up and down trees their automatic recorder is able to measure how much of the journey was up and how much down, and for how long and in what order these phases occurred; then, on the return journey they simply follow the programme through in reverse. The extent to

HUNTING AND SCAVENGING

which these automatic, position-calculating devices are used certainly varies from one species to another; many may well rely entirely on the picture of the area around their nests that they learnt gradually and not use automatic devices at all.

One more important aid to navigation exists and is probably used by all ants: this is a deliberately constructed trail of scent. The ant that has found a lot of food, more than it can take back to the nest in one journey, marks the area by daubing chemicals about. As it returns, it lays a trail of these scent spots all the way back to the nest or track-way. The chemicals, which come from a special gland near the tip of the abdomen, last long enough to be found by other ants but do not sully the area for too long; that would be confusing. When the forager gets back to the nest it runs about excitedly, striking its nest-mates with its antennae and forelegs, and this, with the trail smell as well, stimulates them to run out. Sometimes only one ant comes out and runs so close behind the forager that it frequently touches it; they run in this way, in tandem, as it is called, as far as the food: the first ant is leading the second ant. More often several ants run out alone, without the first; they pick up the scent trail and are able to follow it with care and some hesitation until they reach the food area. There they collect food and return to the nest, marking a trail in their turn. The advantage of a straight run back is now clear: it produces a short direct trail and enables extra helpers to get out to the food as quickly as possible, thus ensuring that the food is collected before any other creature, in particular another colony or species of ant, finds it.

The way ants bring the food in depends on its size and nature. If it is liquid, each sucks up as much as possible in its crop and carries it back a cropful at a time. If it consists of a lot of little solid bits, each ant takes one back; if it consists of a few big solid bits they try to drag and push each piece along towards the nest. Only if this fails do they cut the food up into small pieces, and carry it back gradually. A team of ants struggling with big pieces of food towards the nest looks very inefficient and it is surprising that they get there at all. Presumably each worker knows where it is going, and even though they cannot all pull or push from the best position most of their efforts point in the right direction. Group collection is common with ants.

4. The Collection of Plant Juices and Seeds

In addition to scavenging and hunting, ants collect a great deal of liquid food. Prey or corpses provide them with the materials, called proteins, that are needed to build up their bodies, and they also contain oils for energy. Liquids of course supply water, and usually sugars, that like oil are used for energy, but small amounts of soluble protein are often present as well. Ants suck fluid up through their mouths and pass it down a narrow tube into a special storage reservoir in the abdomen called the crop, which we mentioned in the last chapter. There is a valve between this and the rest of the digestive tract; when closed this stops leakage into the carrier's body, and enables her to regurgitate the food to nest-mates. The crop is, in effect, the public part of the gut and has been called the 'social stomach'. A crop full of liquid distends the worker's abdomen to nearly twice its normal size and ants on a trail that are running back towards the nest are visibly bigger than those going the other way (plate 11). You can easily see this for yourself by putting a bit of thin syrup or honey somewhere near an ants' nest; as the workers find it they will gorge themselves and return home swollen.

The juices that wild ants collect are varied. Sometimes they take plain water: from dew or raindrops, from leaking taps or tanks or just from damp soil. Sometimes they visit the nectaries of flowers, where sugary solutions are secreted and where they accumulate in petal-tubes. Some plants also have open nectaries on their stems and leaves; these are easy to get at.

THE COLLECTION OF PLANT JUICES AND SEEDS

11. The Wood ant (left) gives liquid food to another worker (right). The ant giving liquid has a much bigger gaster (abdomen) than the ant receiving it, since its crop is full of honey-dew. It holds the receiver with its right foreleg whilst its left foreleg stands on a stiff plant thorn. Its antennae lie loosely over the head of the receiving ant and in fact touch it lightly from time to time. The receiver has its antennae folded and retracted and all its legs, except its left foreleg which is poised in mid air, are on the stem. The pedicel scale of the left ant is very conspicuous. Notice too that it has short blunt hairs over most of its body.

Most of the liquids taken by ants come from bugs which suck juices out of plants. These plant-bugs have a set of mouthparts like hollow needles that they push into the soft zones of plants, until sap is forced up into their bodies. They control the flow with a valve, but the plant-bugs usually take far more juice than they need, in order to extract some rather rare materials. This surplus fluid, with some waste substances added from the bug's body, is passed out through their anus. It is then called honey-dew, and contains sugars and a little soluble protein. The exact mixture varies with the part of the plant tapped, the species of plant-bug tapping it and the time of day or year

12. A Black garden worker ant collecting a drop of honey-dew from an aphid. For a short time it holds it on the tip of its tongue. Notice that it stands with its middle and hind legs on the stem and its forelegs on the aphid from which it is collecting liquid.

13. Here an aphid cluster on oak is attended by two Wood ant workers that are touching heads and feeling each other with their antennae. Meanwhile the larva of a hoverfly (white and grub-like) eats one of the aphids without the Wood ants making any attempt to defend it (top aphid in the cluster). The white aphid-like shapes on this picture are skins cast by the aphids after they have moulted and grown into bigger ones.

at which this happens. Most of the plant-bugs used by ants, in this country, are aphids (greenfly) but occasionally coccids (scale insects) are used.

THE COLLECTION OF PLANT JUICES AND SEEDS

Though the use of plant-bugs to tap plant sap has obvious advantages for the ants, the bugs benefit too. They gain mainly by being protected from insect enemies that try to parasitise or eat them, for the ants, with a few exceptions, chase these intruders off. Shelters of soil or vegetable material may be built over the plant-bugs; these protect the whole group, including the attendant ants, from rough weather as well as enemies. Even the plant benefits from the relationship to some extent, for ants, by collecting honey-dew as it is formed, prevent it falling on the leaves and making them sticky; sticky leaves often become covered in a black mould which interferes with their proper functioning. The plant pays slightly for this cleansing, as ants cause bugs to take more sap than they would otherwise do.

Today ants and bugs may have mutal relationships that range from the casual to the completely interdependent. At the casual end of the scale are the aphids that shoot their excretion freely into space; much falls on to the leaves of plants, where ants find it and lick it up. Wasps and bees always collect honey-dew in this way. Other aphids hold their fluids in their hind-gut until ants lightly touch them; only then do they excrete and the ants drink and lick them clean. There is no mess this way. Close contacts of this sort are only possible where the aphids have lost a nasty automatic defensive reaction: that of plastering the ants with a sticky wax from their cornicles. Some ant aphids, as we may call those that regularly associate with ants, have lost their cornicles entirely; as the ants themselves give ample protection from enemies these organs are really unnecessary anyway.

Yellow ants even prepare places for their bugs on plants; they cut the hard outer layers from Dandelion and other roots so that aphids can feed from the soft inner parts. They also carry aphids about from plant to plant and even take them into their nests and care for them as if they were brood. Aphid eggs are collected and stored; if you look inside the mound of a Yellow ant in winter you will be likely to see chambers with many small black eggs in them. In spring these hatch and the young aphid walks to its food plant or, some say, is put on it by the ants. There is an aphid that lives in the nests of the Turf ant (*Tetramorium*), which the workers actually feed by regurgitation. The Turf ant seems to get nothing in return, so that the aphid appears to be a true social parasite; that is, it gets food from the ant as a young cuckoo gets food from a bird of another species such as a hedge-sparrow.

14. Wood ants tending aphids on the young twigs of an oak tree. The ant on the left has its antennae stretched out and is standing with its forefeet on a small aphid. Notice how clearly the head, the thorax and abdomen and the jointed legs stand out.

THE COLLECTION OF PLANT JUICES AND SEEDS

Though this last example shows how aphids sometimes get the better of ants, there is no doubt that usually it is the other way round. Ants have many very curious effects on aphids, not all of which are yet understood. They somehow manage to keep them in dense crowds on the younger parts of the plant and, most puzzling of all, they hinder the aphids from producing winged migratory forms. (These are the aphids that fly away and found new colonies, like winged ants.)

Even more interesting is the fact that some ants culture aphids for their meat, not for their liquids alone, just as we grow cattle for eating as well as milking. The question then is: how do they regulate these two different uses? When do they butcher and when do they milk

15. Workers of the Fungus ants of South America are taking a piece of leaf back to the nest. A small worker rides on the leaf fragment and sucks up the exuding sap, for this contains water and valuable nutrients. Notice how long-legged the workers are.

their aphid stock? Naturalists think that they eat the ones that stray from the group, or the ones that show signs of age, or the ones that get parasitised. They think too that those that fail to deliver honey-dew when palpated, or even worse give a dose of sticky wax, are the ones that are killed. However, other naturalists have provided evidence that coccids are killed when they have become so numerous that the ants are satiated with their secretions. Here is an interesting problem awaiting solution!

Not many ants in this country collect seeds for eating. Wood ants carry them back to the nest to use as nest material and other species nibble seeds before throwing them away. Several species of *Lasius* collect the seeds of gorse and bite off an oily knob called a caruncle. This knob is not essential for the survival of the seed, so that later, when the workers take the seed out and drop it somewhere, the ants are helping the plant to spread itself. If ants throw these seeds in a fairly restricted spot and then add debris so that a rubbish heap is piled up, they are doing something rather like horticulture: sowing seeds and manuring the soil. Ants do not have to understand all the biological processes that are involved, as long as what they do makes the next crop of seeds easier for them to get and more plentiful.

In the American tropics many ants cut leaves from trees (plate 15) and convert them, in their nests, into edible moulds (rather as we cultivate mushrooms). Though rubbish heaps often go mouldy in this country, there is no sign that the ants eat the mould.

5. The Care of Young Ants

Most of the honey-dew collected by foraging ants is immediately shared between the workers in their nest; ants have no special water-tight storage cells like the comb of bees and are forced to consume food as it arrives from outside. Workers are particularly fond of sugar, as you may have found for yourself, and in no time at all they share sweet solutions with each other. The sugar in nectar and honey-dew is a valuable source of energy, and it is energy for movement and building and searching and housework that the adult worker ants need. A little sugar is shared with young stages and with the queens but only if the workers themselves have too much. The prey and seeds collected can be stored: prey for a few days, seeds for a few months.

Like most animals, and nearly all insects, ants lay eggs. These are shaped like a bean and have a smooth, thin, but very tough and resilient, skin made of a pliable material rather like polythene, not rigid and breakable like the shell of a bird's egg. They are only about 100th the size of a worker and are pure white in colour. The small creature that hatches out after a few weeks is grub-like and has no resemblance to the adult ant. It has no legs, it has rings round its body, which is soft, and its head is just a little yellower and harder than the rest of its body. It has two openings: a mouth in front and an anus through which it passes waste materials, behind. Inside, the grub has a straight digestive tube and little else; as it spends its entire life eating and growing, this is almost all it really needs. The head, besides being slightly tougher than the body, has two strongly pointed jaws. As soon as the grub can crawl or wriggle out of its egg it tries to find a soft food-egg laid by workers; it pierces this and sucks off a white, highly nutritious yolky substance.

16. A tropical species of ant closely allied to our Red ant. In the warm parts of the earth ants can live in trees, and this and many other species make their nests in the pith by constructing partitions every few centimetres. Here you can see various stages of brood: eggs, young larvae, old larvae, pupae still in the larval skin, and fully formed pupae (notice their black eyes). Of the last, some are white and some brown and the latter have young adults inside nearly ready to break out. A worker nurse is also present on the top right of the cell.

THE CARE OF YOUNG ANTS

These grubs grow a little after they have had their first meal, but their skin soon becomes much too tight and they have to change into a new one. They make the new skin with sufficient folds and pleats for expansion. After moulting, as the process of getting out of the first skin is called, they start feeding again, but this time they have to get

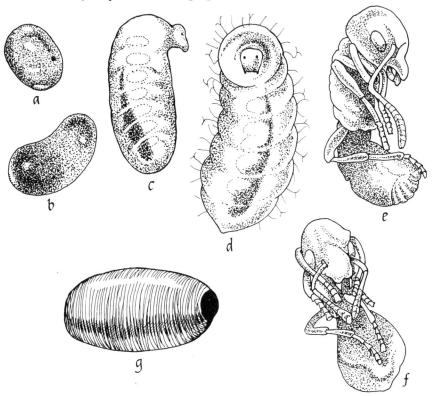

Fig. 5 The brood stages of ants. a. a food egg of the Red ant b. a reproductive egg of the Red ant c. a larva of the Red ant from the side. Notice the head and the dark gut showing through the transparent skin d. another larva of the Red ant seen obliquely from in front: notice the small jaws and the hairs on the back and sides which are short and straight or a little longer and bent with a divided tip e. worker pupa from side: notice how all the main parts of the adult body are formed, including the legs and antennae, but they are wrapped up in pupal skin and the animal cannot move f. a pupa from the front obliquely g. the cocoon of *Lasius* made of a long strand of silk that overlaps itself many times. When this has been spun the larva ejects its food residues, making the black mark on the right, and then changes into a pupa inside.

the food from workers. Workers recognise them, probably because the new skin is hairier than the old one, and pull them out of the cluster of eggs. They are then placed around in a group, lying on their backs with their heads conveniently uppermost, so that the workers can find their mouths and give them food. This is possible because their heads, rather like ours, are bent round towards their under surfaces. The food they are then given is the soluble crop protein that workers are continually digesting and sucking out of prey or seeds.

When nurse workers touch a larva it frequently, though not always, lifts its head up and holds out its mouth. The worker then clasps the head in her two forefeet and places her mouth on the mouth of the larva. Once this contact has been made she forces droplets of liquid food out by telescoping the segments of her abdomen inwards and forwards. The larva sucks the liquid into its gut. Thus the two ants, grub and nurse, co-ordinate their actions and movements so that food is neatly and cleanly transferred from one to the other. Any bits that the larva cannot get are licked up and collected by the worker, so that the whole chamber where brood is nourished does not become messy and liable to invasion by moulds and other noxious organisms.

As a result of feeding on insect juices the guts of the larvae go first brown and then black, and they can be seen quite easily through the clear skin. As the grub grows it moults once more, this time into its third stage, which is even hairier than the previous one. In this stage, which is the last, there are several different kinds of hairs. In addition to the straight pointed ones present in the first and second stages there are many hooked and bent ones. These enable larvae to cling together like little bits of felt so that workers can carry them about in a cluster. Eggs also stick together, but not because they are hairy; they have a thin coating of sticky wax.

Larvae take in enough food to make an adult ant, and then they begin to go through a series of changes which on account of their spectacular character are given a special name: metamorphosis. First of all, the gut wastes, accumulated over the whole life of the larva, are squeezed out; this leaves plenty of room inside. The new stage, called a pupa, separates away from the old skin and so comes to lie in a bath of fluid, a sort of private pool. Legs and antennae grow out and the

THE CARE OF YOUNG ANTS

17. A Black garden ant with cocoons containing large workers. They are spun by the larvae from silk just after they have stopped feeding. After completing the cocoon each larva expels its food residues, which form the black mark visible at one end (the bottom end of the left-hand cocoon).

various body parts take shape in this liquid space. Once the general outline of the adult has been formed the pupa comes out of the old larval skin, starting at the head end through a split down the middle of the back. It inflates its head with body fluids pushed forwards from the abdomen and holds this position for a few moments whilst the soft skin sets firmly.

This creature is still not a real adult, for it cannot walk or fly, but it looks very much like one except that its legs and antennae are folded up; with queens and males even the wings are recognisable as short stumps. Inside, the final touches are being dealt with: a new, more elaborate digestive tube forms, the ovaries grow and form connections so that eggs can get out, and a few other changes take place. Eventually the pupa turns a brown colour and it is possible to see a finished young adult moving about inside. Soon the skin splits, down the back again, and the new ant wriggles out with the help of workers, which pull off the whole skin and take it away. Several weeks are then spent just feeding and resting whilst the skin gradually darkens and toughens; after this the young adult is ready to help with work in the nest.

There is one big difference between myrmicine and formicine ants during metamorphosis. The larvae of the latter first enclose themselves in a silk cocoon spun with material from their salivary glands. However, larvae can only spin successfully against a solid wall and to provide this the workers bury them temporarily in the soil; they dig them up later. Extra help from workers is also needed when the young ant tries to get out of its double envelope. Workers take part in all these intricate changes.

18. A Black garden ant with a cocoon containing a sexual female. It is having difficulty in getting a grip on it preparatory to dragging or carrying it away. Most species of this sub-family (the Formicinae) have larvae that spin cocoons with their saliva just before they change into pupae. In this they differ from the other sub-family (the Myrmicinae).

The life history of ants can now be summarised: from eggs hatch larvae which moult twice, from larvae emerge pupae, and from them, adults. Only the larval stages take food and grow; in the others the structure changes and develops. The whole process from egg to adult takes several months.

Workers spend a lot of time with their brood, particularly with their larvae. The larvae emit from their anus a liquid which contains waste material and is analogous with urine. They give this out whenever a worker lightly touches their hind-body; the nurse then either licks it up and swallows it or carries it away as a droplet in her jaws; this of course keeps the nest clean. Workers then lick around the anus to dry up any residues. They lick around the mouth after feeding too, and also lick and brush the whole surface of brood frequently. There are at least two good reasons for this: first, it removes the spores of bacteria and fungi (like yeasts and mildews) that can invade the brood and perhaps kill it; second, it spreads and polishes the body waxes and so keeps the skin waterproof.

19. Most rubbish and dead ants are taken outside the nest and thrown away, sometimes in a variety of places, sometimes in one place. Here a worker of the Black garden ant is throwing away a dead insect whose soft parts have been eaten.

All the unused bits of food, the cast skins, the old cocoon fragments and the gut residues of larvae are taken out of the nest and dumped some distance away. This collection of refuse is as important for ants as it is for us, for otherwise there is a danger that harmful animals will come to live and breed in it. The chief scavenger is a mite (these are small eight-legged soil animals related to spiders), which lays eggs giving rise to small mite nymphs which climb on to both the adults and the brood. Often so many collect together that proper nursing is difficult and young stages die. Another common scavenger is an eel-worm (nematode); though only 2 to 3 mm long these come in thousands. They too take rides on workers and queens and interfere with work in the colony. These are just two of the many reasons why it is very important for the workers to keep their nests clean.

6. Nests with Ants under Observation

Watching ants in the wild going about their daily work of hunting, collecting food, and carrying it back to the nest is very fascinating and not too difficult. You need to supply food if you are to get a quick reaction and of course you must avoid treading on them. If you want to see the insides of their nests and the way they go about making them, defending them, and dealing with their young much more is needed. Perhaps the best way is to start by trying to look into a natural colony without disturbing it, for although this cannot be completely successful, it is worth trying. The trouble is that ants do not like the light that you need to see them with, so that the only hope is to cover the nest with a transparent red-tinted glass or plastic cover, as they are least sensitive to this colour. Restrict observation to brief spells of a few minutes at a time and use Red ants, which are the most docile type. Cover the glass plate when you are not watching with a tile or slate so that no light is let in and make sure that it cannot slip off. There is no need to make the sheet any bigger than this page, but if you use glass or anything that can be broken accidentally by trampling beasts or men you must protect it; you must be sure too that you have permission to do this, for as you can easily imagine treading on a glass sheet can be very dangerous. If you cannot find the nest easily, try baiting with cake crumbs or bits of insect and watching where the foragers take them. As we have already seen, they go straight back to the nest and it does not take long to find out where

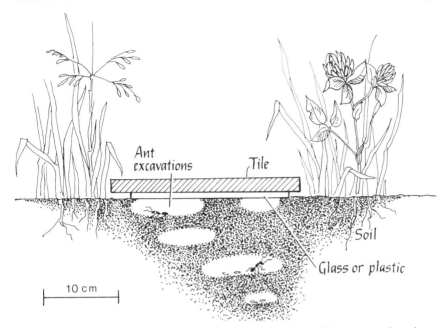

Fig. 6 A diagram of a window on to a natural nest showing a tile lying on a glass sheet under which the ants have excavated chambers

the entrances are. You can then cover these or at least some of them with the tinted sheet.

The ants will build their own earth chambers under the glass or plastic sheet and bring their brood into them, as long as you keep it dark when you are not watching by having the tile or slate firmly in position. You can then watch for different brood stages, for the appearance of pale young adults, and, in summer, for winged sexual ants; you may even see the queens laying eggs. You will be able, with luck, to see something of what they are eating; perhaps you will recognise a centipede or small worm or even a fly. You can of course offer them food too, and watch them take this in and chew it over before eating it.

Watch for daily changes too; do they appear in the morning before you arrive? Do they bring all brood stages up in a mix or are they classified and segregated into eggs, larvae and pupae? How do they build new cells and how does the nest grow? Do the ants stop building once the sexuals have flown and more space is available for the other brood? There are many aspects of interest that can be discovered simply by watching regularly and carefully. You may even be fortunate enough to have another colony of ants edging its way in under the tile and then you can watch how they do this; how they deal with the occupying ant; whether they actually come to grips with each other or simply push in behind a barrier of soil.

The next step in watching a colony is to keep one of your own indoors, which of course demands far more time and attention on your part. The ants behave naturally only if they are well fed and watered, if they are not too hot or too cold and if they have enough darkness and peace. First, you must make the nest container, for this is much more satisfying than having a ready-made one. The simplest nest is a polythene washing-up bowl or a plastic cake container, but any material will do, provided of course, that you can see through it. On the bottom of the bowl put a shallow transparent saucer upside down; it must have a notch in the rim or, if it cannot be cut, a stick underneath so that the ants can come and go (Fig. 7). Make the saucer dark inside by covering it with a piece of black cloth or paper. You will be able to lift this when you want to peer inside. Make the interior moist by putting wet cotton wool or fabric or sponge inside; the ants will use this as a brood chamber. They will come out into the light dry part of the bowl to collect food; and they will, of course, escape over the edge if you do not cover it with a sheet of glass or perspex. Slip a few pieces of paper or cloth under this so that air can move in and out; ants need fresh air all the time. If you can find a substance called Fluon, smear a little of it round the sides of your bowl; after it has dried you will find that the walls are so slippery that the ants cannot climb up. They will fall down to the bottom of the bowl again, and you will not need a cover at all. As an extra precaution against loss and just in case your friends and relations do not like ants running about in the house, you can store the whole bowl in water containing a little detergent.

Of course, if you take ants out of the wild and put them in your nest you have a responsibility to treat them well. Apart from providing a

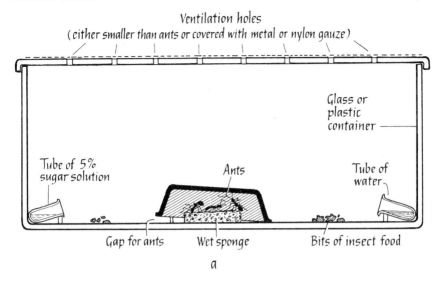
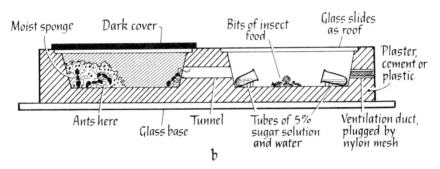

Fig. 7 Two types of artificial nest for indoors. a. an open nest in a plastic container with the lid perforated for ventilation. The ants live in the dark under an inverted opaque saucer and are given food in the light part of the container b. a closed nest made of plaster with shallow chambers covered with sliding glass roofs, one dark in which the ants live and one light where they collect food. There are ventilation holes in the sides.

moist dark chamber inside a light dry foraging area, you must give them fresh water to drink and as much food as they will eat at all times. Water is best given as a very thin sugar solution: one part of sugar (no more) to 20 parts of water. For protein and oil you must give them bits of crushed flies and bits of mealworms or maggots that you can buy from angling or pet shops. Most Red ants do not eat meat or egg, nor even cheese, but other species do. You can try your ants with bits of soft cheese and sweet fruit, but do not leave them to go mouldy. If you want to see ants build nests of their own, give them some moist crumbly soil; as long as it is not too wet they will take it into the dark chamber and build a nest. They like clustering against a wall just as you do in a large uncrowded space; they even like walking on the ceiling (unlike you, I hope). So, if you want to make them really content, let them build their own partitions inside your inverted saucer. To make this easier for them, arrange the ceiling of the chamber (that is, the bottom of the inverted saucer) so that it is a little less than a centimetre from the floor (that is, the bottom of the bowl). As the size of ants varies with species, a good guide is that they should be able just to touch the ceiling with their antennae whilst standing on the floor.

The most difficult thing is to get the ants into your nest, for they will not go quietly. Red ants again are the easiest to collect, mainly because they do not squirt poison into the air space, like some of the formicine species. Even so, they can sting quite badly, and it is best therefore to collect them in winter during mild weather, or in early spring. It may be worth finding a nest in summer and marking it with a stick or a stone; it is not always very easy to find nests in winter, as most of the ants go down into the depths of the soil. Spring is the best time, as the queens are usually in the top of the nest. You can either dig out a huge block of soil with the ants in it and take it quickly to your nest, or excavate the nest gallery by gallery and suck the ants into a tube. The first method works quite well, but you have a great deal of soil to get rid of. This can be done gradually as it dries out, by flaking it away on the outside; the ants will move further and further into the interior of the soil lump and will eventually go under your moist, dark and attractive saucer. You can even connect the bowl containing the soil and ants to your nest bowl by a tube of glass or plastic and let them run in, taking their queens and brood with them. This is a neat way and it causes very little damage to the ants, but it takes time and you have to be very patient.

The other method, excavating a nest, calls for a knife and, of course, a sucking tube. Unless you go carefully, you are likely to crush a lot of the ants and quite possibly they may be able to sting you. If you are lucky you will be able to find a large part of a colony clustered under a stone. You will have to make your sucking tube out of a plastic or thick strong glass tube some 2–3 cm in diameter. Corks that fit in and can be drilled to take plastic or glass tubing of about 5 mm diameter will also be needed. To stop the ants coming out of the tube and yet allow air to pass you will need nylon mesh, which you must tie over the end near your face or fix with adhesive or a rubber band; it is very unpleasant though not dangerous to get ants into your mouth! Flexible rubber or plastic tube of about 30 cm in length is needed to fit over the tubing in the cork, so that you can push your sucking tube down the hole you are excavating whilst keeping your head out. When you have sucked up a colony of ants, including workers and at least one queen, preferably with brood as well, keep it in a cool place before attempting to introduce it to your nest. When the ants are cool and quiet it is quite easy to knock them down into a clump at one end of the sucking tube by means of a few sharp raps. Whilst they are struggling wildly to extricate themselves from this cluster you must take the cork out and, with a few more raps, shake them into your bowl. There will always be a few obstinate individuals, but the majority will go in.

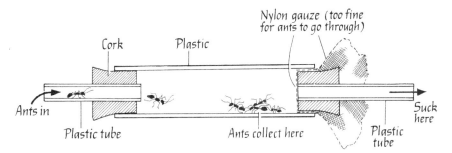

Fig. 8 A sucking tube for collecting ants; it is made of transparent plastic and cork and has a nylon gauze to stop ants being sucked up into your mouth.

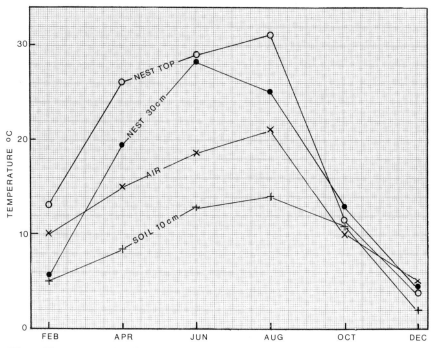

Fig. 9 Temperature in a Wood ants' nest and its surroundings, averaged over a year from many different readings by A. Raignier.
Notice: the nest surface is the warmest in all months except October; 30 cm below the nest surface is warmer than air or soil in spring or summer (April to August). Notice how quickly the nest warms up in spring but loses its advantage in autumn and winter.

Once you have become an expert with these ants you yourself can make and design all sorts of nests, using glass tubes, plastic trays, sliding glass roofs, and even nests made of plaster of Paris with imbedded depressions for chambers. All you need to be sure of is water for drink, some moist darkness for the ants to keep their brood in, and a light dry area for food. Also you will become more and more skilful at letting them make their own nests of soil, or bits of leaf and twig, or even paper or soft plaster; in this way you will learn more about the differences between species. Keep all nests out of the sun at all times, as it gives too great a heat. The best temperature is 20°C; this is the room temperature which most people prefer.

7. Ant Hills as Homes

Ant hills, as we have said, are not just piles of soil or vegetable matter; they are carefully constructed. They are designed to protect the inhabitants from bad weather and their young from the damaging rays of sunlight, and to prevent enemies getting in. They also provide a surface or structure on which to work at rearing new ants (just as your house or school has structures like kitchens or laboratories or benches). Because ants have never learnt how to use tools (as we have) they have no need of many different kinds of workspace. They feel with their antennae and palps, hold with their forefeet and manipulate with their jaws and tongue; quite a simple structure like a room with a low ceiling will do. In this chapter we shall confine ourselves to finding out what sort of shelter an ant hill gives; in other words, what sort of changes it makes in the climate of the soil or area where it lives, what sort of micro-climate (as it is called) is found inside the nest. The word climate is used to include light, temperature, humidity and the atmosphere of the air spaces.

The most important difference between the inside and the outside is that the inside is dark, which means that ants have to feel and smell their way about. They have to recognise each other, and their queens, males and young stages, by shape and smell. Such eyes as they have are no use at all inside the ant hill. However, they are aided in timekeeping since they, like most animals (and plants), almost certainly have a strong daily rhythm; one might almost say that they have a clock inside them. They need to check this against sunrise or sunset from time to time; in between it runs automatically and is even temperature-compensated just as your watch is (that is, it does not run faster in warm weather). This natural timer enables the indoor workers to follow the daily rhythm without continually visiting the outside.

You might well ask why ants do not use sound to communicate with each other in the dark; we certainly would if we could not make artificial light. As you may know, bats use sound to find their way about at night, sending out pulses that come back (that is echo) only if there is an object to reflect them which must therefore be avoided. Curiously enough, some ants, including *Myrmica* species, can make sounds by rubbing a spine on a rilled surface (like a needle on a record). It is even audible to us and I suggest that you hold a *Myrmica* worker just inside your ear; as long as you grip it tightly by the legs (not the body, as this stops it moving its needle), you will hear a chirping noise. The strange thing is that this noise does not affect other ants; although they can probably detect it, it does not seem to carry any message to them. The ant which is making the chirp is usually trapped by a predator or a landslide, yet no ant comes to its help. Ants can detect vibrations in the solid that they are walking on; you only have to touch your observation nest to make them all pause and lift their antennae upwards. They have no ears and probably detect vibrations and pulses by minute movements of hairs on the antennae; this warns them of dangers from large trampling beasts or burrowing moles.

All ants in this country build their nests near the surface of the ground, though most go deep into the soil as well, and many are raised into a hillock well above the normal surface. Bare soil has a different climate from that of the air above. Temperature varies a lot more: it is much hotter when the sun shines but much colder at night, as you must know if you walk about in bare feet at all.

If you have a thermometer that goes over the range $-5°C$ to $60°C$ you can take the soil temperature in a variety of places when the sun shines. You will then discover, for example, how bare soil gets quite hot (30–$40°C$) whereas only a few centimetres of grass is enough to keep the surface cool even in strong sunshine. Try making a pile of

ANT HILLS AS HOMES

loose soil or sand and taking temperatures round it in different places throughout a whole sunny day. Try putting a tile or slate on the ground and see what temperatures you get underneath; thick stones will not get as hot as thin slates. All slabs depend on having an air space underneath them: if they are embedded tightly into the soil they lose heat too quickly to get up a good temperature, for soil, especially if it is wet, conducts heat very quickly and easily. Notice that when cloud hides the sun the temperature stays up for quite a while; if you are lucky enough to have a cloudless day, try shading a slate for some time and then take the shade away. All these things are important to ants in deciding where to make their nest and how to make it.

Before ants start making a nest they select a warm dry spot in a bank that is well drained and faces the morning sun. Morning sun is more important than afternoon sun because it warms the nest up early, whilst the outside air is cool. Yellow ants often build on to their nests on the side facing the morning sun. If this is done year after year the hills come to be shaped, in ground plan, like an ellipse with a long axis east-west and a short axis north-south. You may be able to find nests like this, if you have a compass or can work out the points of the compass from the position of the sun and the time on your watch.

Wood ants' mounds are undoubtedly most cleverly built from the point of view of catching and retaining warmth. Their nest domes are taller in cooler climates than in warmer ones, where the top of the mound can be quite flat. The argument behind this is that a slope which lies directly across the sun's rays, almost at right angles, picks up more heat than a slope which lies obliquely across them; did your soil temperature readings agree with this? In addition Wood ants are said to let the sun's rays penetrate deep into the nest dome by opening up holes in the surface during sunshine and closing them again at night.

Wood ants, Yellow ants and Red ants are exceptionally clever at building mounds; the most that the majority of ants can do is to select a sunny situation and excavate just beneath the soil surface. They usually throw the soil out and scatter it nearby, and it blows or washes away. Some species (for example the Black ant, *Formica fusca*)

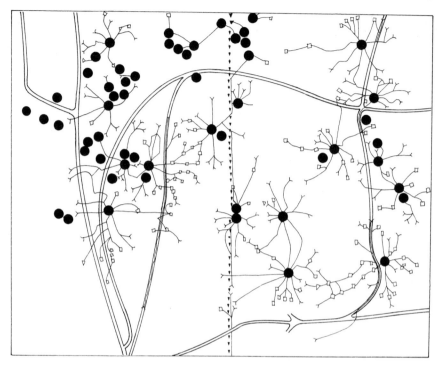

Fig. 10 Nests and foraging tracks of Wood ant colonies observed over eight years in Belgium by A. Raignier. Nests (black circles) send out tracks to trees (white squares) and lead on into the general hunting grounds. The double line tracks are human pathways.

also cut the vegetation down as it grows by nibbling at the place where it leaves the soil surface. Other species have perforce to keep on finding new warm spots in rotting logs, in rotting tree stumps or under stones that have fallen from walls. This makes them very nomadic.

We have said a lot about the ants' search for warmth, but what about their need of moisture? In this country, where rainfall is generally high and the atmosphere and soil moist, most ants avoid water and seek dry places. Raindrops can destroy nests, especially if they are newly made, by just battering them down. In the Lake District, which is notorious for its heavy rains, even the thatched nests of Wood ants can be shattered in this way and workers are constantly rebuilding them. Light rain is very effectively shed, probably because the convex shape of the dome and the general absence of hollows prevents rainwater pooling up and soaking in. No ant, so far as I know, actually thatches with the 'reeds' arranged parallel as we do; they have no way of pinning them down and so cannot achieve the near-perfection of cottage thatch. Nests under stones are well protected from rainfall, but in summer this can be a disadvantage, for the soil right underneath gets too dry; it cracks and becomes too hard for the ants to build with. The only thing they can do then is build their nest round the sides of the stone in the soil which is soft and moist from recent rainfall.

If you look round, you will almost certainly find ants living in very wet places such as marshes and bogs that are sodden with water for much of the year. There is one shiny black ant of the *Formica* type that lives in bogs in the south of this country; it builds a small dome nest of closely woven dead leaves on the top of grass tussocks. Many species of ant are periodically inundated; this usually happens in late winter when it is cold. They then cluster together into a ball, holding air in between their bodies, and this is enough to enable them to survive for several weeks as long as the temperature remains low and they do not respire much. Ants can survive suffocation from gases like carbon dioxide (a product of respiration) for a long time; though fully narcotised, they come round slowly when returning to normal air.

8. Ant Hills through the Year

Ants migrate up and down in the soil as the seasons change. In autumn, well before the cold frosts of Novmber, they filter down their passageways until they reach the lowest galleries of all, right down in the inhospitable subsoil at depths as great as a metre. As the climate above the soil and on the soil surface is cooling rapidly it may well be that they are looking for warmth; they are pursuing the summer heat on its way down into the earth. All the adults go, and if they have larvae, as some do, they take these with them. They pack together much more tightly than when they are actively working. Larvae after they have stopped feeding lose a lot of water and become quite dehydrated; this probably helps them resist cold. Thus the whole colony remains torpid all winter, packed tightly in a cluster well below the surface of the soil. Deep in the soil it never freezes and the temperature does not change a lot: perhaps only by 5 or 10°C. There the ants are free from most enemies; they may become waterlogged, but as we have seen this does not matter at low temperature, because they respire only slightly.

Seed storers do not always take their seeds down with them, neither are aphid eggs taken down by those species that collect them in their nests. Many root-feeding aphids seem to remain in the same place all the year round and it is quite likely that ants lick some honey-dew on warm days in winter. Why not try excavating an ant hill to see just where everything is? As long as it is cold you need not fear being attacked. Try to cut a section, starting at the side and working towards the middle of a nest, but leaving enough of it undamaged for the ants to remake it in spring. Fill in the hole afterwards anyway!

Winter torpor is really only apparent, for inside the ants' bodies a multiplicity of small changes are going on which revitalise them. Inside the queens, new sets of eggs are slowly formed; inside the workers, the digestive glands and the ovaries are reorganised; even the larvae undergo changes that give them a greater power to grow when warmth and food return in spring. This is particularly important for those that stand a chance of growing into huge queen-forming grubs. Thus the winter break restores ant bodies; the temperature is low enough to make feeding unnecessary, but high enough for important chemical changes to go on slowly.

In spring, when the top of the nest begins to warm up, the workers gather in the new heat; they usually cluster just under the soil crust, but Wood ants come right out into the sunlight. The workers spend a lot of time at first just sunning themselves; then they repair the nest or build up the mound. They also have to eject soil animals that have invaded the surface galleries during winter. Such animals are slugs, earthworms, centipedes, earwigs and woodlice, all of which like to rest by day in the shelter of unused ant nests. The queens soon join the workers, and they too take some time to come to life—to reactivate their organs in the warmth. Several weeks may elapse before foraging starts and larvae are brought up, rehydrated and fed.

Often a month or more of larval growth goes on in quite cool conditions before the queens lay many eggs. In the Red ant (*Myrmica*) temperatures must exceed 15°C before they can regularly lay fertile eggs, though larvae can grow at temperatures of 8°C or more. The best temperature, 20°C, is the one the Red ants choose and the one at which they grow and survive best.

By May big colonies show their huge sexual larvae. In most myrmicine ants these are formed from big winter larvae (those which were taken into the ground for the winter), but in Wood ants it is different; they come from eggs laid very early in the season, at the top of the hill, by the queens. After these come a series of eggs that give worker-forming larvae, but these are laid, after a few weeks' pause, in the deeper parts of the nest.

The next significant event is metamorphosis. This definitely needs a higher temperature than either larval growth or egg laying, for in cool May weather it can be held up, only to burst out again when warmth returns. Sometimes, in *Myrmica* all the winter larvae start to metamorphose before the eggs hatch. There is then a short period, of a few weeks perhaps, when the nests contain no larvae at all. This gives the young workers, still quite fresh from their winter's rest, a chance to lay large numbers of eggs. These are rather flabby, never develop into larvae and are used as food. Workers lay more of these food eggs if the queens themselves are also in full lay; in fact, the queens eat food eggs, thereby increasing their own egg production. Many food eggs are put in the same pile as the queen eggs, where they remain until young larvae hatch. These grub-like creatures only have to wriggle a little distance in order to find a food egg, which they pierce and suck dry. Just one of these eggs makes a substantial meal for a young larva and gives it a good start in life, for later it may have to compete with other larvae for the attention of workers.

Egg-laying is an extraordinarily good way for workers to occupy themselves whilst waiting for larvae to hatch. In a normal nest with queens they produce only these flabby edible eggs. When the queens stop laying or die, the workers start to lay big thick-skinned eggs like those of the queens, and these give rise to males.

Shortly after the queens' eggs have begun to hatch, new adults emerge from pupae, giving first young queens and males, and then workers. This usually starts in early June, and it is obviously a good arrangement that new larvae and the new workers needed to look after them should appear together. It is usually some weeks before new workers are ready to nurse, for they first eat food themselves to build up their own internal stores. Only the males go without food after they emerge from their pupal skin (or cocoon in formicine ants); young queens eat and store a lot.

The flight of sexuals from the nest (see Chapter 9) leaves a lot more space and probably more peace as well. Afterwards the new young workers can concentrate on rearing brood and feeding the queen so that as many eggs as possible can be produced for the formation of winter larvae. The time when new workers take over brood-rearing is a very significant one, for larvae that are reared by these new nurses, though they grow well, do not metamorphose; instead they become dormant and before long dehydrate ready for hibernation. The dehydrated larva is the only brood stage that is able to survive the cold of winter. The arrival of the new workers in the nest thus starts preparation for winter. Since no more pupae are formed, their number declines as they give rise to more workers; also the queen soon stops laying, partly because the workers do not prepare food for her and partly because by then she has laid all the eggs she can. As a result it is not long before there are neither eggs nor pupae left in the nest, and the colony, well before winter, comes to consist only of adults and larvae.

This preparation must start in good time even in summer, because it takes a month or so for eggs to hatch and for pupae to change into adults. After that the internal reserves of the adults must be fully charged and the larvae must grow before being dehydrated. When all this is complete the colony can move down to the lowest possible level in the soil well before the first frosts of winter arrive.

9. Winged Ants and Nuptial Flights

The winged sexual ants emerge from their pupal skins, and though the females feed and store up reserves whilst their skin hardens and darkens, the males eat very little. Only after these preliminaries are the sexuals fit to go out of the ant hill into sunshine, rain and wind. Only after building up their stores are the queens able to sustain themselves, without any food at all, for long enough to make workers. Naturally, to increase their chances of success they stay in the nest until the weather is favourable: that is, calm, warm and moist.

Then a curious thing happens: after weeks in the darkness of the ant hill they all go out into the light together, in a tumultuous rush away from their home nest. The whole colony is affected: even workers open holes in the roof and rush out. The sexuals climb up grass and herbs or on to twigs, open their wings and fly off: not a vigorous flight like that of a bee but a flight just strong enough to carry them up and away against light winds. These massive flights occur over wide areas of the country at the same time and naturalists have, not suprisingly, speculated a great deal about how this happens. Obviously one factor is that the weather can be suitable over wide geographical areas; but it is also curious that each species seems to fly at a particular time of day. Thus the Turf ant (*Tetramorium*) flies in the early morning, the Black garden ant (*Lasius niger*) about mid-day and the Yellow ant (*L. flavus*) in the late afternoon. Now it is known that many insects can sense time of day: they have, in effect, internal clocks. Ant sexuals in fact rise to the surface repeatedly at a particular

20. A male Wood ant ready to fly off for its nuptials. Notice its extended antennae, probably sensing the humidity and chemistry of the air. It has large eyes and a large thorax full of muscles for vibrating its wings, which are at the moment folded over its back. The scale on the petiole is clearly visible and at the tip of its abdomen are the claspers which interlock with the female's genitalia during copulation.

time for several days before they actually take off. Each species does this at a characteristic hour, and it is this time sense coupled with a weather sense that enables flying ants to emerge simultaneously over wide regions of the earth.

Different species of ant mature their sexuals and have nuptial flights in different months: for example, the Turf ant (*Tetramorium*) has ripe sexuals in its nests quite early in June, whereas most Red ants (*Myrmica*) do not have them until mid-July. The Turf ant queen cannot overwinter alone; she needs the company of workers. So this species has to get new queens early enough in the year for them to lay eggs and mature them into workers. This of course prevents *Tetramorium* from living in the north of this country; they can only manage to produce new young workers in warm places in the south. *Myrmica* on the other hand do not need workers for their first hibernation; they can survive quite well alone or just in groups of queens. This may be an important asset and enable Red ants to live further north than the Turf ant.

Even the species of *Lasius* differ. Though the Yellow ant and the Black ant (*L. flavus* and *L. niger* respectively) fly in the period mid-July to mid-August, another species called *L. alienus* (rather like *L. niger*) does not fly until September or even October. In the British Isles good weather may be over by then, and in a cold wet autumn the sexuals never get a chance to fly and are all killed and eaten by their workers. In the Mediterranean zone, where summer can be too hot and dry, October is often ideal and it is in these climates that *L. alienus* prospers.

You may wonder just why so many ant nuptial flights happen at the same time. The basic answer is that species whose individuals mate with those of a different nest fairly frequently are more likely to produce new forms that can survive in our changing, fluctuating, unstable and highly competitive world. Out-breeding, or cross-mating, as it may be called, favours the evolution of adaptability in the species. So a queen from a colony here is more likely to have high quality offspring if she mates with a male from another colony ten miles away. To do this she must take off at the same time on the same day, though this alone does not guarantee success. These massive flights attract enemies to the feast that flying ants provide. The first ant to emerge may well avoid most enemies, but later ones can only hope to survive if they pour out in such a mass that the predators are overwhelmed by the great mass of ants and soon become satiated. Alternatively the sexuals can sneak out surreptitiously and sit, hidden in a tree, emitting a powerful scent that attracts the opposite sex (this is what the Wood ant females do). Of the two methods the former is more costly—more ants die—but it is more likely to achieve substantial cross-mating and so to mix the hereditary stock and produce new types of individual that can equip the species better for survival.

After they have left the nest the sexuals usually congregate together in special mating places. Species of *Myrmica* congregate over trees, outstanding rocks or houses, or even at the peak of a mountain. If the wind blows a faint gust they drop momentarily down to the sheltered side. In fact, they swarm in almost the same way as many flies, such as midges, do, using their eyes to fix a position near a suitable landmark. The males form the main swarms; females only come to find males and then to pair. They are about the same size as the males, so both drop to the ground and complete sperm transfer there; this usually takes place on a bare patch of ground or a roof, near or under the marker tree or rock.

Only species of *Lasius* seem to manage to bring opposite sexes together without a special meeting place; they just fly up into the air. It is quite likely, in spite of appearances to the contrary, that they do not go up beyond a distance that enables them to detect and follow some sort of pattern on the ground. They may tend to fly over light patches like pathways. They may also fly over areas where convection currents are rising, such as houses and fields, rather than where they are falling, such as woods. If you watch them with field glasses (be careful not to look at the sun, for this can damage your eyes) you may see them drifting by in clusters that look as though there is a queen in the centre surrounded by a flock or crowd of males. *Lasius* males are much smaller than their females and a female can carry a male in flight whilst they are actually pairing. This takes only a few minutes, after which the male drops off and the female goes in search of a place in which to start her nest.

After pairing with one male (perhaps several) the female has a full store of sperm and can dispense with males for the rest of her life. Her next problem is equally urgent: she must find a good situation. This may be a place in which to start a new colony, or a colony of her species that will accept her as an auxiliary queen, or even a colony that needs a replacement for its existing queen. You would expect her to hunt for the right sort of general habitat, say grassland or scrub or woodland, whilst still airborne, for clearly she could explore much more ground in a short time this way. Perhaps this happens, but we are not sure. One piece of evidence is that Black ant queens (*Lasius niger*) will land on bare garden soil rather than soil covered in vegetables; presumably they can detect the greater daytime warmth of the bare soil, but they may be repelled by the smell of cabbages! Queens of the same species also land on moist heath rather than dry heath, but whether this shows that they can sense humidity from the air or not is doubtful. The Turf ant (*Tetramorium*), in contrast, lands on dry heath, and so does the ant called *Lasius alienus*, whose home it is.

Once the ant queen has found the right sort of place (habitat), it is still vitally important that she finds a bare spot of soil or wood that the sun will warm. She may do this before or after she has broken her wings off—we do not know—but a guess is that she more often does it afterwards. Unless a queen finds a warm spot she will not be able to rear a brood of young workers either in the same season (as the Turf ant does) or in the next one. Though established colonies can build mounds of soil in plants and cut them down, queens cannot; they are too weak and helpless. Even a little vegetation spreading over the nest site, as you may have found in your temperature experiments, can make it much cooler.

Her last act is to dig into the soil or wood and excavate a cell; if the weather is too dry this will be difficult for her. Hence the importance of flying after a shower of rain has softened the earth. It is a fact of life that most of the best ant places will already be full of ants. Moreover, they will usually be hostile to newcomers. In spite of the dangers, some species of ant make a point of locating already inhabited areas and trying to get into an established colony of their own species or even into one of another species. Wood ant queens do this. They are rarely found starting their own colonies; more often they try to enter another Wood ant colony or a colony of the Black ant (*Formica fusca*), and eliminate the resident queen or make the workers eliminate her. The Black ant workers are then used to bring on the new queen's first brood of Wood ant workers. Later, of course, the Black host workers, as they are called, die out completely, but as the Wood ant has got over the critical settling-down stage with their help this does not matter (to the Wood ant!). This is called temporary social parasitism and is reminiscent of the behaviour of cuckoos. The fact that Wood ant queens can recognise the right species of host shows that their sense of smell must be quite well developed, and leads us to think that other ants may be just as well equipped.

Many queens are often found in a single ant colony. There are several possible causes: they may have settled together and dug a common nest, or some may have joined after the colony was a few years old. Sometimes these new queens are daughters of the original foundress, but this is not always so.

10. The Foundation of Colonies and their Growth

Queens that have found a good place in which to settle are, as we have just said, quite likely to find that others have got there first or come along soon afterwards. The better the place and the richer the area in ants, the more likely is this to happen. Very sensibly, instead of squabbling they dig a hole together and make a common cell at the end of it, and by living peaceably in this way they can pool their resources and increase their chances of success. They do not go out to hunt food at all, but instead make eggs out of body stores and liquidised wing muscles which in the absence of wings are no longer required. They can guard their nest much more effectively as a group than they could alone. When the first eggs hatch the larvae are fed on other eggs; they grow, and quickly change into new workers which are normally smaller than average. They break open the seals of the nest and wander out to collect fresh food. Of our ants only the Turf ant achieves workers in the same summer as the nuptials; others may produce a few larvae but these normally change into workers only in the following spring.

The appearance of the first new workers is a critical event in the development of the colony in many types of ant. Queens often become intolerant of each other; perhaps they vie for the attention of the workers. They fight, and make their own individual cells, and then try to collect brood into these and guard it from their erstwhile companions. In some species like the Black ant (*Lasius niger*) so much hostility develops that many queens are killed. In others, like the

21. A huge queen, length several centimetres, of a fungus-culturing ant from South America. These queens found their colonies alone like many of our species but start to cultivate a fungus garden with spores carried in their mouth pockets. Then they and the new small workers (seen climbing over the queen) feed on the fungus. Eventually the workers go out and collect pieces of leaf which are first cleaned and chewed and then given to the fungus garden to convert into edible 'mushroom'. This ant does not live in cool parts of the world.

THE FOUNDATION OF COLONIES AND THEIR GROWTH

Yellow ant (*Lasius flavus*), a dominant queen is usually able to monopolise the brood and continue the colony alone, without the help of the others. In Red ants (species of *Myrmica*) this crisis never arises or at least has no serious consequences; colonies can grow to full size with many queens and then add more!

The nuptial season is a good one in which to start observation nests. You can also make original observations on the extent to which queens co-operate. In July or August if you wander about outside you are almost bound to encounter young queens running on the soil or on paths and roads. Collect them up and put them in groups in your nest. Until workers appear they will need only water on a damp sponge. They are best kept cool (5 to 10°C) until the spring, so that their bodies are prepared for egg laying. Then give them a little sugar and keep them at 20°C; as soon as workers hatch provide them with all the normal foods, including of course the all-important growth food: protein.

Starting with queens and new workers, the colony grows because it can bring up more new workers each year than the number of old ones that die, provided of course that it is in a good place and well fed. So more and more workers accumulate, even if only a few of them live more than two years. As the colony grows it reaches out further to get food and the workers lay food eggs to help the queen. This sounds fine, but there is a snag: the expenditure of more energy and time in hunting means that the food begins to 'cost' more. Bigger workers are formed, and since these can travel further with more efficiency they can bring in more food. But this helps only for a time, as there is a limit to the size of workers. The colony will then cease to grow, and the workers will collect as much food as they can as near the nest as possible.

Before this stage is reached the colony is likely to meet a neighbour. What happens then will be the subject of the next chapter; here we can consider the other ways in which ants start new colonies. They can bud from parent colonies, or a colony can divide into two parts which go in different directions. The latter method is rare in this country, but Wood ants have sometimes been seen dividing in early spring. In forming a bud workers go out and prospect for a good situation. They virtually sample it by living there for a while and testing its temperature variation, food supply and general suitability. If they find it attractive, they fetch brood and eggs, and in due course even queens. The transport of queens is not really necessary, for they can be made from eggs as long as enough of these, laid by the parent queen, have been taken to the new nest site. The larvae that hatch can be specially nourished, and the workers can lay their own male-producing eggs. Then the two sexes can pair in the nest and start the bud off as a self-sufficient and independent entity.

The first sign of maturity in a colony with big workers is the production of a few males. In some ants these males come from eggs laid by workers (this is probably the case in most *Myrmica*). In others they come from eggs laid by the queens (probably in most Wood ants). Queen eggs that give males must be unfertilised, because the entry of a sperm makes the egg female.* Very little is known about what stops the sperm entering these eggs; the sperm store may be empty or the queens may have started to lay early in spring, before the weather is warm enough for the sperm to penetrate the eggs. Fertilisation may fail if the queens are disturbed by the constant jostling of crowded workers. If queens are getting old, their organs may cease to function for many reasons. As we have seen, eggs from workers are able to develop only if their queens are not laying or do not have sufficient contact with their workers. This may happen if they are old or if the colony is too congested. Then the workers cease to produce their food-eggs and start laying reproductive eggs which are very like those of queens in shape and firmness, but a trifle bigger. These develop into male larvae and yield the first adult males. Thus the production of males by a growing colony starts when the queen fails to lay sufficient eggs, when her eggs are unfertilised, and when she no longer has adequate contact with her workers.

If female sexuals are to be produced, workers must be unable to encounter their queens at all or at least rarely; even a dead queen can stop workers making new queens. In big nests that spread widely,

---

*All ants, bees and wasps, unlike most animals, produce males from unfertilised and females from fertilised eggs.

since the queens cluster together and do not walk around on patrol, workers can quite easily escape their controlling influence. The queens do not emit a scent or a sound that could travel, and workers have to keep literally in touch with them nearly all the time. Otherwise their behaviour changes. They neglect small larvae with a tendency to produce workers and feed big ones with a disposition to grow into queens; they do this lavishly. These larvae grow then at a fantastic speed, reach a huge size and metamorphose into young sexual females. Most ants do not stop at producing a few replacements. They produce vast quantities; enough for export as well as for re-stocking!

I have described what happens in Red ants. In the Wood ants the process is very similar; queens come to the warm nest tops in spring and lay their eggs. After a few days they go down and rest deep in the nest. The eggs, still on the surface, hatch into larvae and are reared by workers that are sufficiently out of touch with their virtually subterranean queens to behave as though they had none. This means that they bring the larvae up on a rich diet and mature sexual females. Males are produced at the same time, it is thought from eggs laid by the queens at too low a temperature for fertilisation to occur. Later on, in summer, the queens start laying again, but this time they remain deep in the nest. In the warmer summer weather, for some reason that we don't yet understand, they can no longer lay queen-disposed eggs and the second brood consists entirely of workers.

There is an advantage in holding back the production of sexuals. They do a colony no good; they do not work or help in any way, and are, in effect, parasites. Hence it is essential that the colony should be strong enough to bear them without suffering in the process. Suppression of sexuals by the queens ensures that the colony does not form them prematurely. You can think of the queens as opposing a worker pressure to make sexual forms. Suddenly their opposition collapses and the colony matures with a burst of adults of both sexes.

11. The Pattern of Ant Communities

Most colonies try to expand in spring. Wood ants, in their search for prey and for trees with aphids, often encounter other colonies at this time of year. Then a battle starts and workers from both colonies try to drag each other back to their own nest. New workers are brought up to the battle zone in just the same way as when food is discovered. Fighting diminishes at night but is resumed in the same place in the morning. Another species that fights in spring is the Turf ant (*Tetramorium*). These little black ants hang on to each other by the legs so that hundreds can often be seen gripped tightly together in a complicated mass; they seem to be trying to drag each other back to their nests. I do not think it is by any means certain that all these affrays really involve lethal fighting; many of them do very little damage to the workers that take part. They should probably be thought of as contests or trials of strength between neighbouring colonies. A strong colony with a nest near the contest zone, and linked with it by a good transport and communication system, has every advantage and will probably drive the other away.

The outcome, whatever the savagery involved, is that colonies establish the sole power to forage over areas of land or around clusters of aphids. In the case of the Wood ant whole trees with their aphid fauna can be monopolised. It is rare, for example, for two colonies of Wood ants to share the same tree; when they do, one goes up one side of the trunk and the other up the other side, so that their trails and trackways never cross. Occasionally trails cross by means of bridges

THE PATTERN OF ANT COMMUNITIES 75

formed by chance sticks or fallen boughs. Hence each colony in spring develops a network of track-ways that never join or contact those of other colonies; they mark out an area, including trees and bushes, which is defended by the ants. As with birds, it is called a territory. Turf ants and Yellow ants have underground galleries that enable them to maintain their territories. Red ants do not have clear track-ways at all; they simply cover and guard clusters of aphids on herbs a few metres away from their nest. Usually, because the traffic between the clusters and the nest is heavy, and the risk of hostile encounters high, workers from other colonies avoid these areas.

The establishment of territories has the effect of spacing the ants out over the available ground. You can easily see that the mounds of the Yellow ant are spread out over whole hillsides, not as regularly as are our houses but so well spaced that two are rarely very close together. Ecologists, in fact, have shown by measuring their distance apart that this spacing is something between what they call 'random' and what they call 'regular'. Why not try measuring the distances between ant hills? It is simplest to measure the distance from each hill to the nearest neighbouring hill. Then group these measurements according to whether they are, say, less than 5 metres, between 5 and 10 metres, between 10 and 15 metres and so on. If you then plot a diagram representing the numbers in each of these classes you will obtain what statisticians call a frequency distribution. What shape is it?

Where more than one species of ant tries to live in an area, complications arise. The outcome depends on whether the place in which they meet is intricate enough for them to share it, and at the same time avoid friction. One species must live in one part and the other in another part, or they must live in different ways. For example, one may live in wet and the other in dry soil; one may live high up in trees and the other down in herbs; or they may nest quite close together but feed in different places or on different foods, one perhaps on seeds and the other on prey. If the place is very simple and uniform this sort of sharing will not be possible and only one will survive. It is remarkable how species do manage to share a habitat, so we will look now at a few examples.

In the west of Scotland two species of Red ant occur, *Myrmica ruginodis* and *M. scabrinodis*: call them R and S respectively. There is not a lot of warmth in these latitudes and the summer season is quite short. Both Red ants like grassland and heather moor such as grows on the sides of mountains and they struggle for the sunny places. S almost always win; their queens can settle and oust the queens of R and their colonies can surround, besiege, and drive out those of R. In fact, R usually move rather than wait to be destroyed one at a time. Once S have taken over a warm spot they hang on to it better than R would. They make a strong thick-walled nest of soil containing many small rooms with partitions of plastered soil; such a nest not only keeps out enemies but resists drying and desiccation as well.

R, once evicted, go to live in taller heather, nesting in moss in the centre of the old bushes, or into much longer grass that is scarcely grazed by sheep and that rabbits very often do not find attractive; here they nest under fallen wall stones. They may even settle in scrub or woodland where there are glades that get a few hours of sunshine; in such places they are away from all competition by other species of ant. They survive because they are able to build a special light-weight tent-like nest which reaches up to the sunshine and captures heat. The tents are quite unlike the soil nests of S; they have an outer wall composed of vegetation and soil, interlaced with natural herb structures, and they contain relatively large rooms. In fact, they look like an open-plan house rather than a conventional one with many small rooms. Such light, easily constructed nests do not involve any great investment of labour or material on the part of R and they can readily move when the growth of shrubs and trees, or even of herbs, deprives them of the sunshine they need. Their simple, easily made but effective nests enable them to be very mobile, in fact almost nomadic, and to live in a habitat in which the sun may shine here one year and there the next.

R have at least one other feature that enables them to live among the foliage of quite tall herbs. Their legs are long, so they can stretch from one leaf or stem to another and travel quite quickly, like a monkey, some ten centimetres or so above the ground surface. S with their short legs and crouching habit are much more at home moving on the soil between the stem bases of plants and squeezing through narrow soil apertures and surface irregularities.

As a matter of fact, in very short vegetation which allows a lot of sunlight to reach the soil, the warmest level is just above the surface. In longer vegetation which screens the soil from direct sunlight the hottest level is usually some distance above the ground, depending upon the nature of the vegetation; a common height would be 5 to 10 centimetres and this is where R live. Thus each species forages and builds its nest in the level or stratum which is hottest for that particular type of habitat, that is, height of grass.

If we look at a slightly more complicated habitat, we find, in addition to R and S, two more ant species. This is old woodland, cleared of trees and left to regenerate naturally. The tree stumps, the only remains of trees cut down five or ten years previously, are rotting, and between these grass and heather grow, with here and there young birch and willow trees. These small trees provide a new layer for the ants to forage in and there is one, a Black ant (*Formica lemani*), which ascends them in search of honey-dew. It nests in bare spaces on the ground, in particular on the sunny side of large rotting tree stumps. *F. lemani* is a powerful, active and very aggressive ant and can easily drive both of the two Red species away. They are usually forced to nest to the north of the stump, in the cooler part of the soft wood, or even away in the longer herbs. However, the Black ant, though powerful, lacks the building skill of the Red ones and just dies out when the trees grow and shade its nest place.

The fourth species is a small brown myrmicine ant of the genus *Leptothorax*. In spite of being small and peaceful in its habits, it survives in the presence of the other three because it has a special skill that the others lack. This is the ability to cut nest galleries and chambers in quite hard wood; it makes the galleries so narrow that the bigger ants cannot get in. This species also has very small colonies which can manage to live in hollowed-out twigs and acorns. Thus there is no competition for nest sites between it and the others. Its diet is unknown but cannot conflict seriously with that of the other ants. So, the presence of bushes and trees and rotting tree stumps in heath and grassland enables two extra species of ant to live there.

It is interesting to compare these ants of the Scottish moor and grassland with ants living in heather in the south of England on sandy soils ranging from Surrey to Dorset. Instead of four species there are nearer ten, and though the two Red ants we have been talking about are there, they play only a minor role. The Black ant present belongs to the species *Formica fusca* but is very similar to the Scottish *Formica lemani*, except as regards its nesting habits: it digs deep nests with a single entrance. This last feature enables it to guard the nest by posting a group of workers round the entrance hole, a method which is so effective that this species can live in the best parts of the heath surrounded by other ants. The brown *Leptothorax* is also replaced by a slightly different species. To these northern forms are added the Black garden ant (*Lasius niger*) a frequent inhabitant, as its name implies, of English gardens. There are also several other species which would prefer to live in warmer climates: a brown one called *Lasius alienus* and the Turf ant *Tetramorium caespitum* that we have already mentioned several times. There is even a predominantly Mediterranean species called *Tapinoma erraticum*.

The three principal species are the Black garden ant, the brown *Lasius alienus* and the Turf ant. We can look at the way they fit together here as we did with the ants in Scottish heaths. The Black garden ant, call it N, can be found in the wet grassy heaths, where it nests in tussocks of grass and any clump of vegetation which stands up above the water level. This ant feeds on various plant-sucking bugs and also by catching prey, mostly above ground. The workers are comparatively large and strong and can capture quite big insects like honey-bees and grasshoppers. Their shiny surface and black pigmentation enable them to forage in bright sunshine and their hairiness enables them to fall in water without risk of drowning. N queens actually choose moist places in preference to dry ones in heathland.

The other species, *Lasius alienus*, call it A, inhabits much drier and more windswept areas, where the plants are short and patchy and where much of the soil is bare or covered in small scrubby lichens. To live in these arid windswept places it makes underground foraging galleries and hunts soil animals; it also rears aphids on various heath plants that are enclosed and covered in soil. The workers of this *Lasius* are smaller and weaker than those of N, but this enables them to penetrate into soil crevices more easily in their hunt for small insects. They do not need a dark or shiny skin if they venture above ground only in dull weather, and as they live in dry places they run no risk of drowning and can dispense with hairs that might impede their movement underground. The queens, unlike those of N, select dry places in preference to wet ones.

You would hardly expect the two species to meet at all, but they do, and then N treats A very badly indeed. N workers collect around the entrances to A nests and frequently dive in, grab and pull out A workers, which they take away and dismember. In this way, though they do not succeed in driving A away altogether, they do succeed in keeping them underground, with the result that the two ants become neatly layered: N above, A below, the soil surface. This is similar but not quite the same as the relationship between the two species S and R in Scotland. S keep R out of short vegetation but do not live below R in long vegetation; they never become layered. In both cases the aggressive species lack skills possessed by the others: for example, the ability of A to make foraging galleries in the soil, or the ability of R to make tent nests and move quickly through herbs.

In the southern heathland the best established species is the Turf ant; it lives in the richest, most productive positions, the ones that are warm, moist and well vegetated. It has the power to push either of the two *Lasius* species (N or A) on one side, for it combines the ability to make underground galleries (as can A) with the ability to forage in herbs (as does N), and adds to these the capacity for a high degree of co-operation between its workers (both in hunting and fighting) and the ability to live for a large part of the year on seeds. It stocks seeds, after collecting them in autumn, in underground chambers, and consumes them in spring. It makes full use of the varied heathland resources, and because of this and its aggressive dominance it usually inhabits a zone between the two *Lasius* species, pushing A into the more arid and N into the more humid parts of the heath, incidentally reducing the friction between them.

12. Plants and Animals that use Ant Hills

There are quite a number of plants that are characteristic of the hills made by Yellow ants. Many of these are annual plants that grow each year from seed, living only long enough to produce more seed and then dying and vanishing into the soil. Plants like this find it difficult to grow amongst vegetation that is already well established. In the shorter sparser vegetation of growing ant hills they find it an easy matter to establish their seeds in loose freshly brought-up soil on the top of the mound. Examples are the Chickweeds, Sandworts, Speedwells and some annual grasses. Other plants are there before the ants start building and are not suffocated or swamped when soil is piled on top of them; in fact, they have a special ability to send up new shoots through the soil. This keeps their whole shoot system growing inside the ant hill from a few original roots beneath it. They form an important part of the structural girder system which gives mechanical stability and strength to the ant hill. Examples are the wild Thyme, the wild Rockrose and several grasses such as Bent and Fescue.

These Yellow ant hills attract rabbits, which sit on the platform at the top. Whilst sitting they produce a lot of dung in pellet form; this certainly benefits the plants underneath and hence the aphids that feed on their roots, and in turn the ants which collect honey-dew from the aphids. In this way rabbits and Yellow ants are mutually beneficial. In winter, though, rabbits may dig into the hills.

Since the soil of these hillocks is light, friable and warm, it attracts

22. This hill of the Yellow ant shows the effect of rabbits and woodpeckers. They scratch into the sides so that soil is loosened and the roots of plants are exposed. In spring the Yellow ants repair this damage.

23. A rabbit has scratched into the side of this Yellow ant hill. Loose soil has been thrown out and the bare roots of surface plants exposed.

24. During winter, Wood ants retreat to the deep soil below the ground and the pile of pine needles and other vegetable debris is broken up and distributed by rabbits and woodpeckers which dig into the surface and scatter it.

many insects. These feed on grasses and lay their eggs near the surface. Then there is the Click-beetle, whose larvae are long, thin rather tough creatures called wireworms; they feed on the roots of some plants and are in turn eaten by the Yellow ants. Grasshoppers also come to the surface of the mound, feed on the vegetation, and lay their eggs in the soft surface soil.

Wood ants do not tolerate plants on their hills; any that appear are cut down or have their bark chewed off until they die. Mammals and birds can only do damage in winter. Yet quite a number of insects have evolved subtle methods of entering. There is a moth that lays its eggs at the nest entrance whilst keeping the ants at bay with a chemical repellent; grubs hatch, build themselves a protective case of silk, interwoven with nest fabric, and then start to feed on the vegetable material of the mound. There is a Chafer-beetle which lays its eggs in the nests and whose larvae eat nest material; they avoid the worker ants by burrowing quickly when touched and by having tough bristly skins. There is another beetle, called *Clythra*, whose larvae actually eat ant mouth-pellets. The adult beetle drops eggs disguised with a smear of its own faeces from bushes above the mound, and the worker ants, unable to distinguish them from nest material, take them inside, where they yield larvae which make themselves protective cases and continue to live safely in the nest galleries.

Many beetles have evolved a closer relationship with ants. Though most only eat ants casually, as they would any insect they met whilst hunting, some are able to enter nest galleries and catch workers, particularly those that stray away from their fellows. Any ant that shows fight is subdued by a smear of repellent from glands in the beetle's abdomen. Beetles in the nest can of course reach the ant brood chambers and find the queens and their eggs. One seems to have achieved a pleasant and easy way of eating eggs; it stands on the queen's back whilst she is laying and takes them as they appear, perfectly fresh. In addition these beetles eat other brood stages and are even fed on ant food by the workers, who take a secretion from the beetle's abdomen in exchange.

A notorious beetle, called *Atemeles*, lives with Wood ants. Both insects pile eggs together into mixed clusters, but the beetle eggs hatch first and its young feed on the eggs that are left, including its own! These beetles also offer gifts to the worker ants. They are remarkable in that the young adult leaves the Wood ant colony in autumn and goes to live with Red ants over winter. Instead of walking into a hostile reception it waits near the entrance of a Red ant nest and offers gifts to foragers from glands down the sides of its body until eventually it is picked up and taken in. The beetle still has two emergency devices in case of trouble: one is a substance which subdues and pacifies aggressive Red ants; the other is a repellent and is only used in extreme emergency. This beetle, then, has a complete kit for entering Red ant nests!

There is also a group of butterflies called 'Blues' whose caterpillars

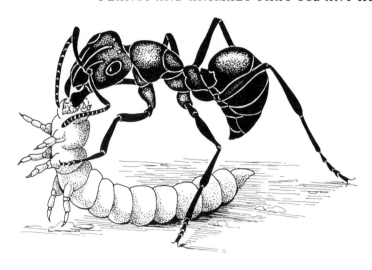

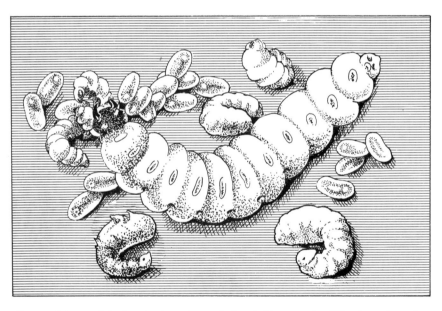

Fig. 11 A larva of the beetle *Atemeles*. a. being fed by a worker Wood ant b. eating small larvae of the Wood ant in its nest chambers.

PLANTS AND ANIMALS THAT USE ANT HILLS

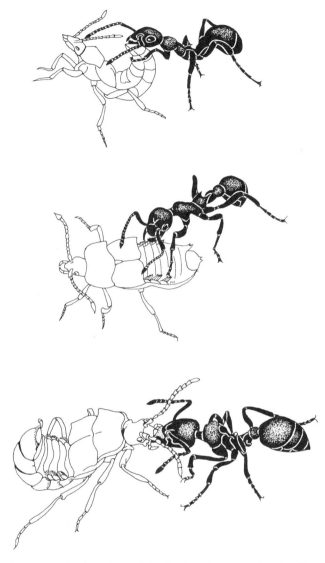

Fig. 12 Adult of beetle *Atemeles* and the Red ant host. a. a beetle offers a worker a taste of its tip gland and so reduces the worker's hostility b. the ant then senses the beetle's side glands that cause it to lift the beetle and carry it into the nest c. liquid food is regurgitated by the ant to the beetle.

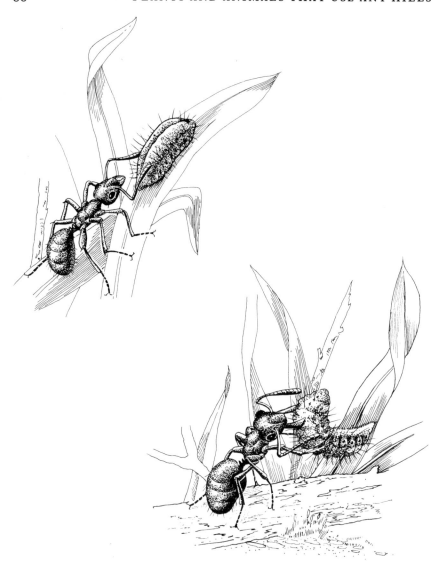

Fig. 13 Caterpillars of the Large blue butterfly, and ants.
The Large blue caterpillar is being recognised by the Red ant. a. one discovers and licks the attractive secretion from its hind end b. the larva then rears up and the ant begins to carry it towards its nest.

are very attractive to ants. The Chalk hill blue caterpillar feeds on Horseshoe vetch and may be surrounded by crowds of the Yellow ant; workers may even protect it with soil covers. This is as far as the relationship goes, but another butterfly called the Large blue has become completely dependent on ants. Its caterpillar feeds on the flowers of Thyme, and when it is found by Red ants a group of foragers collect round it but do not build a shelter: they go one better and take it into their nest for the winter. However, this is a fatal step, for once inside the nest the butterfly larva changes its habits and eats the ant brood. Thus it not only gains full protection from enemies and winter weather but has a magnificent food supply too.

Some kinds of ant go into other ants' nests for shelter, thus saving themselves the time and trouble of making their own nest. Several small myrmicine ants actually live with the big Wood ants in this way and they can be seen running about unmolested on the surface of the mounds; they are small and agile enough to avoid any hostile contact. They obtain shelter and warmth and perhaps some food stolen from the droplets produced by Wood ant workers as they exchange liquids. Many of these myrmicine ants make nests of their own inside the nest of bigger ants, not only Wood ants but Yellow ants and the various Red ants too. Do not forget to look for them when you excavate a colony. Some ants share the whole nest structure so that workers of both species run about in the same galleries. These mixed societies begin with the guest queen entering an established colony of the host, usually a small one; there she either kills the host queen or gains the allegiance of the host workers so that they do it for her; she manages to make herself more attractive than the real queen. The death of the host queen inevitably means that the host colony eventually dies out, so the guest has the use of it for only a short time and must then either die out in its turn or become independent. Such temporary social parasitism, as it is called, is very common in ants: for example, Wood ants can start new colonies by means of queens that enter the nests of the allied Black ant (*Formica fusca*).

There is also a group of species belonging to the genus *Lasius* that use this device of temporary social parasitism to start new colonies. The common Black garden ant (*Lasius niger*) is the host for a brown species (*L. umbratus*), whose queens have a curious way of entering small host colonies when they find them. They first make a cell near the entrance and wait until they can pick up a worker as it goes by. They eat it and lose all their aggressive tendencies. This enables them to walk unopposed into the host's nest, for it seems that as long as they themselves behave quietly and submissively they are safe from attack. Eventually the workers come to prefer the new queen to their own, who is sooner or later killed, and they rear the parasite young that hatch. This cannot last, of course, and when the last Black worker dies the brown *Lasius umbratus* are on their own.

Social parasitic ants have managed to prolong their period of dependence on a host in two ways: either they go out and collect more workers from other colonies of the host species or they refrain from killing the host queen and rely on getting her under control. The first method is practised by an ant that occurs in a few places in this country, the Red slavemaking ant (*Formica sanguinea*). Superficially this species is very like a Wood ant (*F. rufa*) both as regards appearance and habits and, like them, its queen, after nuptials, searches for a nest of the Black ant (*Formica fusca*), enters it and takes over from the Black queen. Her progeny grow and replace the Black workers, but the *F. sanguinea* workers are poor nurses and nothing like as good with larvae as the Black ones. They compensate for this by being much better hunters, and they spend their time scouting out nests of the Black ant; then, after assembling together, they set up a mass raid during which they collect brood and take it back to their own nest. Much of this is eaten, just as if it were any other sort of insect, but not all; out of the pupae that survive new Black workers emerge. They are adopted and they soon help the aged and dwindling originals; for this reason they have been called 'slaves'.

Those social parasites that refrain from killing their host queens are somehow able to prevent her producing sexual brood. She continues to give rise to workers, which nourish the parasite species. In many cases, in fact, the parasite workers are quite unable to forage and feed themselves and have become totally dependent on their host. Such a case occurs in the south of England: the host is the Turf ant and the parasite is a similar and related ant whose jaws are no longer strong and toothed like those of the Turf ant, but weak and pointed. Some ant social parasites produce no workers of their own at all and have come to rely entirely on those of their hosts. The Turf ant is also parasitised by such a species.

Further Reading

J. H. Sudd. *An Introduction to the Behaviour of Ants.* Arnold, London 1967
R. Chauvin. *The World of Ants.* Gollancz, London 1970 (translated from the French by G. Ordish)
M. V. Brian. *Ants. New Naturalist No. 59.* Collins, London 1977

Index

abdomen, 24, 26, 28; Fig. 1; Pls. 5, 14, 20
anatomy, 20; Fig. 1
antennae, 20, 24, 30, 44, 55, 56; Figs. 1, 2, 5; Pls. 5, 7, 8, 9, 11, 12, 13, 20
anus, 24, 41, 46; Figs. 1, 3
aphid, 29, 35, 37, 60, 74, 75, 78, 80; Pls. 7, 8, 12, 13, 14
Atemeles, 83; Figs. 11, 12

bait, 29, 34, 48
bee, 26, 28, 37, 41, 64, 72, 78
beetle, 29, 82, 83; Figs. 11, 12
bent, 80
birch, 77
Black ant (*Formica fusca*), 57, 68, 69, 78, 87, 88
Black garden ant (*Lasius niger*), 18, 64, 67, 78, 87; Fig. 4; Pls. 8, 12, 17, 18, 19
Blue butterfly, 83, 87; Fig. 13

caterpillars, 29, 87; Fig. 13
Centipede, 29, 61
Chafer, 83
Chalkhill blue butterfly, 87
Chickweed, 80
Click-beetle, 82
Clythra, 83
coccid, 36, 40
cocoon, 46, 62; Fig. 5; Pls. 17, 18
collecting, 19, 52
colony, development, 71; Pl. 21
 division, 71
 foundation, 26, 68, 69, 87

competition, 69, 74, 75

Dandelion, 37
defence mechanism, 24, 28, 29, 82, 83; Fig. 12
diet, 29, 31, 34, 39, 40, 41, 42, 44, 51, 60, 62, 64, 78, 79, 82, 87, 88; Pls. 10, 15, 21
downland, 12
Driver ant, Pl. 6

Earthworm, 29, 61; Pl. 10
Earwig, 61
Eelworm, 47
eggs, 24, 26, 28, 41, 44, 61, 62, 65, 69, 72; Fig. 5; Pl. 16
emmet, 13
excretion, 35, 37, 44, 46; Fig. 5; Pl. 17
eyes, 20, 26; Fig. 2; Pls. 5, 7, 20

Fescue, 80
flies, 29
fluon, 50
food, *see* diet
food, collection, 31, 71; Pls. 7, 8, 10, 12, 13, 14, 21
 sharing, 34, 38, 41, 44, 87; Fig. 12; Pl. 11
 transport, 33; Pls. 10, 15
foraging, 30, 34, 74
formic acid, 16
Formica, 28, 59
Formica fusca, *see* Black ant
Formica lemani, 77, 78
Formica rufa, *see* Wood ant

Formica sanguinea, 88
Formicidae, 28
Formicinae, 28, 46, 62; Fig. 4; Pl. 18
froghopper, 29
Fungus ant, Pls. 15, 21

gorse, 40
grasshopper, 29, 78
greenfly, *see* aphid
gut, 20, 24, 41, 44, 45; Figs. 3, 5

habitat, 11, 12, 15, 18, 19, 67, 75, 76, 77, 78
hair, 44, 56, 78, 79; Fig. 5; Pl. 11
head, 20, 26, 41; Figs. 1, 2, 3; Pls. 5, 14
hibernation, 61, 63, 65
honeydew, 35, 37, 40, 41, 77, 80; Pl. 12 (*see also* aphid)
Horseshoe vetch, 87
hoverfly, Pl. 12
humidity, 55; Fig. 20 (*see also* moisture)
hygiene, 37, 44, 46, 47, 61; Pls. 8, 9, 19

internal clock, 32, 55, 64

jaws, 13, 20, 30, 31, 42, 55, 89; Figs. 1, 2; Pls. 7, 10

Large blue butterfly, 87; Fig. 13
larva, 41, 46, 60, 61, 62, 63, 73; Figs. 5, 11; Pls. 16, 17
Lasius, 28, 40, 67, 79, 87; Fig. 5
L. alienus, 65, 67, 78
L. flavus, *see* Yellow ant
L. niger, *see* Black garden ant
L. umbratus, 87, 88
legs, 13, 24, 44, 76; Figs. 1, 5; Pls. 5, 7, 8, 11, 12, 13, 14, 15
Leptothorax, 77, 78
life history, 46, 61; Fig. 5

mating, 26, 64, 66; Pl. 20
males, 26, 45, 62, 66, 72; Fig. 4; Pl. 20
metamorphosis, 44, 46, 62
mite, 47

moisture, 14, 34, 50, 54, 58, 60, 63, 64, 71, 78 (*see also* humidity)
mole, 11, 56
moth, 29, 82
mould, 40 (*see also* Fungus ant)
moult, 43, 44
Myrmica, 28, 46, 56, 62, 66, 71, 72 (*see also* Red ant)
M. ruginodis, 76
M. scabrinodis, 76
Myrmicinae, 28, 46, 61, 87; Fig. 4; Pl. 17

navigation, 32
nectary, 34
nervous system, 20; Fig. 3
nest, artificial, 50, 71; Fig. 7
  dimensions, 13, 14, 16
  material, 11, 13, 14, 15, 77; Pls. 4, 5, 6
  observation, 48, 71; Fig. 6
  site, 11, 57, 76; Pls. 1, 2
  structure, 11, 16, 18, 19, 55, 57, 59, 61, 76, 77, 78, 80; Pls. 1, 2, 3, 4, 5, 6, 16, 22, 23, 24
nuptial flight, 64, 65
nursing, 41, 44, 46, 62, 88; Pl. 16
nutrition, *see* diet

ovary, 24, 26, 45

palp, 20, 40, 55
parasite, 38, 68, 73, 87
pasture, 11; Pl. 1
pedicel, 28; Pl. 11 (*see also* petiole)
petiole, Fig. 4; Pl. 20 (*see also* pedicel)
plant-bug, 29, 35, 36, 37
poison apparatus, 24, 28, 30; Fig. 3
predator, 26, 31, 66, 83, 87; Figs. 11, 12, 13; Pl. 22
prey, 29, 31, 39, 41, 78, 82, 88; Pl. 10
projects, 11, 17, 19, 20, 26, 48, 50, 60, 71, 87
pupa, 44, 45, 62; Fig. 5; Pls. 16, 17

queen, 25, 45, 52, 61, 62, 65, 66, 68, 69, 73, 87; Fig. 4; Pl. 21

INDEX

rabbit, 76, 80; Pls. 22, 23, 24
Red ant (*Myrmica* sp.), 19, 48, 52, 57, 61, 65, 71, 73, 75, 76, 77, 78, 83, 87; Figs. 3, 5, 12, 13; Pl. 16
Red slavemaking ant, 88
Rockrose, 80
rubbish, 40, 47; Pl. 19

Sandwort, 80
scale insect, 35
scent, 33, 66, 73
Scottish moorland, 76
seasonal variation, 25, 60, 65, 73, 74; Fig. 9; Pl. 24
sexual production, 61, 62, 64, 72, 73, 88; Pl. 18
skin, 44, 45, 47; Fig. 5
slug, 29, 61
soil, 12, 13, 14, 18, 19, 52, 56, 57, 59, 60, 76, 78, 79, 80
sound, 56, 73
southern heathland, 78
Speedwell, 80
sperm, 26, 66, 67, 72
spider, 29
springtail, 29
sting, 19, 28, 30; Fig. 3
stones, 11, 58, 76
sucking tube, 53; Fig. 8
sunlight, 11, 16, 32, 55, 57, 76; Pl. 3
swarm, 66

Tapinoma erraticum, 78
temperature, 54, 55, 56, 57, 60, 61, 67, 71, 76; Fig. 9
Tetramorium caespitum, *see* Turf ant
thorax, 24, 26; Fig. 1; Pls. 5, 14, 20
Thyme, 80, 87
trail, 16, 31, 33, 74; Fig. 10
tree stump, 11, 16, 58, 77
Turf ant (*Tetramorium caespitum*), 38, 64, 65, 67, 69, 74, 75, 78, 79, 88; Fig. 4

vibration, 56

wasp, 26, 28, 37, 72
weather, 14, 16, 37, 58, 64, 78; Fig. 9; Pl. 4
willow, 77
wings, 24, 26, 45, 68; Pl. 20
winter larva, 61, 62, 63
wireworms, 29, 82
Wood ant (*Formica rufa*), 12, 15, 40, 57, 59, 61, 66, 68, 71, 72, 73, 74, 82, 83, 87, 88; Figs. 1, 3, 9, 11; Pls. 3, 4, 5, 7, 9, 10, 11, 13, 14, 20, 24
woodland, 12, 15, 77; Pl. 3
woodlice, 61
woodpecker, Pls. 22, 24

Yellow ant (*Lasius flavus*), 12, 13, 14, 16, 37, 57, 64, 71, 75, 80, 82, 87; Pls. 1, 2, 22, 23